1 Mathematical models in mechanics

1.1 Mathematical modelling in mechanics

The basic ideas of mathematical modelling as applied in mechanics.

Mathematical models can be used in mechanics to make predictions about real-life situations. Models must be tested and improved by comparing their predictions against experimental data, but once they successfully describe the situation they can then be used to describe many similar situations.

In order to develop a model you must simplify the real-life situation by making **modelling assumptions.** For example, if you are trying to model a car that travels 5 km, you may ask where you treat as the starting point of measurement. Do you measure from the front end of the car or the back end? Because the car is small relative to the distance travelled, you may assume that the car can be treated as a particle. This then becomes a modelling assumption. Similarly, the time taken for the car to travel may be influenced by wind; it will reach a given destination faster if there is a strong wind travelling in its direction of travel. In M1 you assume that any effects due to wind can be neglected. There are several terms and definitions used in M1 that you should know (you may be examined on these types of modelling assumptions):

Objects treated as particles: in general, it is assumed that objects that are small relative to the other sizes involved will be treated as particles; this means that the mass of the object can be considered to act at the single point where the particle is placed, e.g. an aeroplane can be modelled as a point particle, as it travels long distances compared to its length.

Lamina: this is a flat object whose thickness can be ignored, as the thickness is small compared to the other dimensions (its width and its length). For example, a thin sheet of card may be represented as a lamina.

Rigid body: this is an object that is made up of different parts or masses which do not move relative to each other. For example, a car can be considered to be a rigid body, i.e. when the engine produces a forward force, the whole car moves.

Rod: this is an object in which the mass is considered to concentrate along a line; it is assumed to have only length, and not breadth nor width.

Uniform object (lamina or rod): if an object is uniform then the mass is evenly distributed across the object and can be considered to act at its centre. For example, you could make the assumption that a tree trunk is a uniform rod and that its mass can be considered to act from its centre. However, this may not be a reasonable modelling assumption if the base of the trunk is thicker than the top of the trunk. In this case, the trunk is non-uniform and the centre of mass may have to be calculated.

Light object: when the mass of an object is small compared to the rest of the masses involved, it is said to be a light object. For example, when modelling the mass of the Earth, the mass of a car is relatively light, so the mass of the car may be ignored in the model.

Light and inextensible strings: in M1, strings are assumed to be:

1 light, i.e. the mass of the string is small relative to the other masses involved. This is a fair assumption, unless you are using a chain instead of a string, in which case the mass may have to be included.

2 inextensible, i.e. you don't have to account for any force to extend the string. Again, this is a fair assumption, as the extension in a string is relatively small.

More complicated problems, without these assumptions, are introduced in M3.

Smooth surface: when a surface is assumed to be smooth then you assume that there is no resistive force opposing the movement due to the contact of the surface with the object, i.e. you assume that there is no **friction**. For example, when modelling an ice rink, you may assume that the surface is smooth.

Rough surface: when the surface is not smooth then it is said to be rough. In this case the friction must be included when modelling the situation. For example, sand paper being rubbed across a brick will have a comparatively large force of friction that cannot be ignored because the contact between the surfaces is rough.

Smooth and light pulleys: in M1, pulleys are assumed to be:

1 smooth, i.e. there is no friction on the surface or the bearings of the pulley.

2 light, i.e. the mass of the pulley is considered to be small in comparison to the other masses involved.

Bead: this is a particle that can be threaded onto a string or wire.

Wire: this is a rigid body that is a thin thread of metal.

Peg: this is a support from which an object can be hung or on which an object may rest. It acts as a point but may be either rough or smooth.

Gravity: this is the force that attracts objects towards each other. Because of the relatively massive size of the Earth, it can be assumed that objects only experience attraction towards the Earth and not towards each other. The force of gravity reduces in size as you move further away from the surface of the Earth. However, in all the models in M1, gravity is assumed to be constant.

Air resistance: this is a force that is experienced because of the resistance of the air. For example, when you drop a sheet of paper, the paper will not fall as fast as a brick. However, in M1 it is generally assumed that there is no air resistance.

Wind: this is a force that can be felt because of the action of the wind. In M1 it is assumed that there is no wind.

Advanced Maths Es
Mechanics 1 for ED

CD ROM
assisted book
CD Acc: 106610

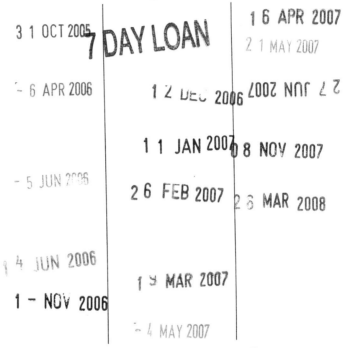
Welcome to Advanced Maths Essen
you to improve your exam performa
need in your EDEXCEL Mechanics

The book contains scores of worked
problem. You can then apply the step
exam questions at the end of each ch
you can try the Skills Check Extra e
this book there is a practice exam-st

Some of the questions in the book ha
PowerPoint® solution (on the CD-R
the problem and setting out your ans

Using the CD-ROM

To use the accompanying CD-ROM
should appear automatically. If it doesn't automatically run on your PC:

1. Select the My Computer icon on your desktop.
2. Select the CD-ROM drive icon.
3. Select Open.
4. Select mechanics1_for _edexcel.exe

If you don't have PowerPoint® on your computer you can download PowerPoint 2003
Viewer®. This will allow you to view and print the presentations. Download the viewer from
http://www.microsoft.com

Pearson Education Limited
Edinburgh Gate
Harlow
Essex
CM20 2JE
England
www.longman.co.uk

First published 2004
ISBN   0 582 83675 8

Design by Ken Vail Graphic Design

Cover design by Raven Design

Typeset by Tech-Set, Gateshead

Printed in the U.K. by Scotprint, Haddington

The publisher's policy is to use paper manufactured from sustainable forests.

The Publisher wishes to draw attention to the Single-User Licence Agreement situated at the back of the book. Please read this agreement carefully before installing and using the CD-ROM.

We are grateful for permission from London Qualifications Ltd., trading as Edexcel to reproduce past exam questions. All such questions have a reference in the margin. London Qualifications Ltd., trading as Edexcel can accept no responsibility whatsoever for accuracy of any solutions or answers to these questions.

Every effort has been made to ensure that the structure and level of sample question papers matches the current specification requirements and that solutions are accurate. However, the publisher can accept no responsibility whatsoever for accuracy of any solutions or answers to these questions. Any such solutions or answers may not necessarily constitute all possible solutions.

2 Vectors in mechanics

2.1 Vectors

Magnitude and direction of a vector. Resultant of vectors may also be required.

A **vector quantity** is one that has both size and direction, as opposed to a **scalar quantity** that has size only. When you add two scalar quantities, you can add the numbers directly. When you add two vector quantities, you must take their directions into account first.

Example 2.1 A man walks 2 km due north from point O to point A and then 3 km due east to point B. Find the distance OB and the bearing of the point B from O.

Step 1: Draw a clear diagram, marking all known angles and distances.

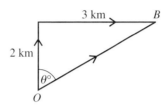

Recall:
Bearings are measured clockwise from north (in this case the line OA).

Step 2: Use trigonometry to calculate unknown angles and distances.

$$OB^2 = OA^2 + AB^2$$
$$= 2^2 + 3^2 = 13$$
$$OB = \sqrt{13} \text{ km}$$
$$\tan \theta° = 3/2$$
$$\theta = 56.3\ldots$$

The bearing is 056° (nearest degree).

Tip:
OAB is a right-angled triangle. Solve using Pythagoras' theorem.

Example 2.2 The point A is on a bearing of 045° and at a distance of 25 m from the point O. The point B is on a bearing of 160° and at a distance of 40 m from the point A. Find the distance OB and the bearing of B from O.

Step 1: Draw a clear diagram, marking all known angles and distances.

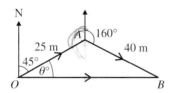

Angle $OAB = 360° - 135° - 160°$
$$= 65°$$
Let angle AOB be θ.

Recall:
Interior angles on parallel lines add up to 180° and angles around a point add up to 360°.

Step 2: Use trigonometry to calculate unknown angles and distances.

$$OB^2 = OA^2 + AB^2 - 2(OA)(AB) \cos 65°$$
$$= 25^2 + 40^2 - 2(25)(40) \cos 65° = 1379.76\ldots$$
$$OB = 37.1 \text{ m (3 s.f.)}$$

$$\frac{\sin \theta°}{AB} = \frac{\sin 65°}{OB}$$

$$\sin \theta° = \frac{40 \sin 65°}{37.1\ldots} = 0.9759\ldots$$

$$\theta = 77.4\ldots$$

The bearing of B from O is $(45° + 77.4°) = 122°$ (nearest degree).

Tip:
OAB is not a right-angled triangle. Find OB using the cosine rule.
Find θ using the sine rule.
The sine and cosine rules are covered in C2.

Notation: In the above example the **distance** OA is 25 m, while the **vector** OA is 25 m on a bearing of 045°. To distinguish between the distance OA and the vector OA an arrow is placed over the letters to represent the vector thus: $\overrightarrow{OA}$.

Note:
The direction of the arrow over the letters describes the direction of the vector.

Example 2.2 can be described in this notation as:

$$\overrightarrow{OB} = \overrightarrow{OA} + \overrightarrow{AB}$$

The vector $\overrightarrow{OB}$ is called the **resultant** of the vectors $\overrightarrow{OA}$ and $\overrightarrow{AB}$.

Notation: You can also represent vectors using a single letter in bold type. If the vector $\overrightarrow{OA} = \mathbf{x}$, the vector $\overrightarrow{AB} = \mathbf{y}$, and the vector $\overrightarrow{OB} = \mathbf{z}$, then:

$$\mathbf{z} = \mathbf{x} + \mathbf{y}.$$

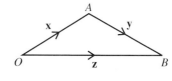

Note:
If the vector from O to $A = \mathbf{x}$, then the vector from A to $O = -\mathbf{x}$
i.e. $\overrightarrow{OA} = \mathbf{x}$
$\overrightarrow{AO} = -\mathbf{x}$

The size of the vector $\mathbf{x}$ can be written as $|\mathbf{x}|$ or simply x without the bold type. This refers to the **scalar magnitude**: the distance OA.

If two vectors, $\mathbf{x}$ and $\mathbf{y}$, are equal then you can write:

$$\mathbf{x} = \mathbf{y}$$

Note:
This means that they both have the same size and direction.

If two vectors, $\mathbf{x}$ and $\mathbf{y}$, are parallel but have different magnitudes, then you can write:

$$\mathbf{x} = k\mathbf{y},$$

where k is a scalar.

Note:
This means that they both have the same direction, but not necessarily the same size: the vector $\mathbf{x}$ is k times bigger than the vector $\mathbf{y}$.

When you handwrite vectors, since you cannot show bold type, you underline the letter, for example <u>a</u> or a̲.

Example 2.3 $OABC$ is a parallelogram with $\overrightarrow{OA} = \mathbf{a}$ and $\overrightarrow{OC} = \mathbf{c}$. The point M is the midpoint of BC and the point N is on AB such that $AN = \frac{2}{3}AB$.

Express, in terms of $\mathbf{a}$ and $\mathbf{c}$, the following vectors:

a $\overrightarrow{AB}$ **b** $\overrightarrow{OB}$ **c** $\overrightarrow{AO}$ **d** $\overrightarrow{BO}$

e $\overrightarrow{OM}$ **f** $\overrightarrow{CN}$ **g** $\overrightarrow{MN}$

Step 1: Draw a clear diagram, marking all known vectors.

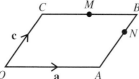

Note:
Because $OABC$ is a parallelogram, BC is parallel to OA and so $\overrightarrow{CB} = \mathbf{a}$ and similarly $\overrightarrow{AB} = \mathbf{c}$.

a $\overrightarrow{AB} = \mathbf{c}$

b $\overrightarrow{OB} = \overrightarrow{OA} + \overrightarrow{AB} = \mathbf{a} + \mathbf{c}$

c $\overrightarrow{AO} = -\mathbf{a}$

d $\overrightarrow{BO} = -\overrightarrow{OB} = -\mathbf{a} - \mathbf{c}$

e $\overrightarrow{OM} = \overrightarrow{OC} + \overrightarrow{CM}$

$\qquad = \overrightarrow{OC} + \frac{1}{2}\overrightarrow{CB}$

$\qquad = \mathbf{c} + \frac{1}{2}\mathbf{a}$

f $\overrightarrow{CN} = \overrightarrow{CB} + \overrightarrow{BN}$

$\qquad = \overrightarrow{CB} + \frac{1}{3}\overrightarrow{BA}$

$\qquad = \mathbf{a} - \frac{1}{3}\mathbf{c}$

g $\overrightarrow{MN} = \overrightarrow{MB} + \overrightarrow{BN}$

$\qquad = \frac{1}{2}\mathbf{a} - \frac{1}{3}\mathbf{c}$

Note:
(c) $\overrightarrow{AO}$ is in the opposite direction to $\overrightarrow{OA}$.

Note:
(f) You can also calculate $\overrightarrow{CN} = \overrightarrow{CO} + \overrightarrow{OA} + \overrightarrow{AN}$ to give the same result.

The i and j vectors

Note:
The $-3\mathbf{j}$ means that you move 3 units in the negative direction along the y-axis.

Vector **i** represents one unit in the positive direction along the x-axis.
Vector **j** represents one unit in the positive direction along the y-axis.

For example, the vector $2\mathbf{i} - 3\mathbf{j}$
from the origin can be drawn as:

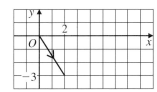

The size and direction of the vector $a\mathbf{i} + b\mathbf{j}$

Note:
The size of a vector can also be called the **magnitude** or **modulus**.

If the vector $\mathbf{v} = a\mathbf{i} + b\mathbf{j}$, then you can find the size of $\mathbf{v}$, written $|\mathbf{v}|$ or v, by using Pythagoras' theorem.

$$|\mathbf{v}| = \sqrt{a^2 + b^2}$$

and $\tan \theta^\circ = \dfrac{b}{a}$

where θ° is the angle that the vector
makes with the positive x-axis in an anticlockwise direction. Use a
diagram to find the acute angle made with the x-axis and then find the
angle made with the positive x-axis.

Example 2.4   Calculate the modulus, $|\mathbf{v}|$, and the angle anticlockwise from the
positive x-axis when:

Step 1: Draw a clear
diagram, marking all
known angles and
distances.

a $\mathbf{v} = 2\mathbf{i} + 4\mathbf{j}$

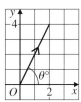

b $\mathbf{v} = 3\mathbf{i} - \mathbf{j}$

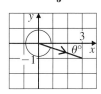

Step 2: Use trigonometry
to calculate unknown
angles and distances.

$|\mathbf{v}| = \sqrt{2^2 + 4^2} = 4.47$ (3 s.f.)

$\tan \theta^\circ = \frac{4}{2}$

$\quad \theta = 63.4\ldots$

The angle between the vector
and the positive x-axis is
63.4° (3 s.f.)

$|\mathbf{v}| = \sqrt{3^2 + (-1)^2} = 3.16$ (3 s.f.)

$\tan \theta^\circ = \frac{1}{3}$

$\quad \theta = 18.4\ldots$

The angle between the vector
and the positive x-axis is
$360° - 18.4\ldots° = 342°$ (3 s.f.)

Recall:
$|\mathbf{v}| = \sqrt{a^2 + b^2}$

Note:
Imagine the vector makes a
right-angled triangle with the
axis, solve for the acute angle,
then calculate the angle anticlockwise from the x-axis.

The i and j components of any vector

A vector of modulus $|\mathbf{M}|$, or M, that acts at an angle θ to the positive x-axis can also be expressed in the form $x\mathbf{i} + y\mathbf{j}$. You can calculate these horizontal and vertical **components** using simple trigonometry:

The horizontal component, $x\mathbf{i}$:

$\cos \theta^\circ = \dfrac{x}{M}$

$x = M \cos \theta^\circ$

The vertical component, $y\mathbf{j}$:

$\sin \theta^\circ = \dfrac{y}{M}$

$y = M \sin \theta$

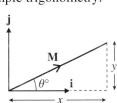

$$\boxed{x\mathbf{i} + y\mathbf{j} = M \cos \theta^\circ \mathbf{i} + M \sin \theta^\circ \mathbf{j}}$$

Example 2.5 **a** The vector OA has a magnitude of 4 units and acts at an angle of 30° to the vector **i**. Find the vector OA in the form $x\mathbf{i} + y\mathbf{j}$.

b The vector OB has magnitude 6 units and acts at an angle of 130° to the vector **i**. Find the vector OB in the form $x\mathbf{i} + y\mathbf{j}$.

Note:
Let O be at the origin.

Step 1: Draw a clear diagram, marking all known angles and distances.

a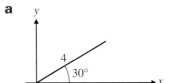

b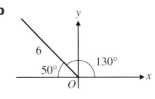

Step 2: Use trigonometry to calculate unknown angles and distances.

$$x\mathbf{i} + y\mathbf{j} = M \cos \theta \mathbf{i} + M \sin \theta \mathbf{j}$$
$$= 4 \cos 30°\mathbf{i} + 4 \sin 30°\mathbf{j}$$
$$= 2\sqrt{3} + 2\mathbf{j}$$

$$x\mathbf{i} + y\mathbf{j} = -6 \cos (50°)\mathbf{i} + 6 \sin (50°)\mathbf{j}$$
$$= -3.86\mathbf{i} + 4.60\mathbf{j} \text{ (3 s.f.)}$$

Tip:
First calculate the acute angle made with the axis then indicate the direction as $+$ve or $-$ve.

Adding and subtracting vectors

$$(a_1\mathbf{i} + a_2\mathbf{j}) + (b_1\mathbf{i} + b_2\mathbf{j}) = (a_1 + b_1)\mathbf{i} + (a_2 + b_2)\mathbf{j}$$

You add the **i** components and the **j** components separately.

$$(a_1\mathbf{i} + a_2\mathbf{j}) - (b_1\mathbf{i} + b_2\mathbf{j}) = (a_1 - b_1)\mathbf{i} + (a_2 - b_2)\mathbf{j}$$

You subtract the **i** components and the **j** components separately.

Multiplying a vector by a constant

When k is a constant, then

$$k(a_1\mathbf{i} + a_2\mathbf{j}) = ka_1\mathbf{i} + ka_2\mathbf{j}$$

You multiply the **i** component by k and the **j** component by k.

Example 2.6 Find the following in the form $x\mathbf{i} + y\mathbf{j}$ when $\mathbf{a} = 2\mathbf{i} + 4\mathbf{j}$ and $\mathbf{b} = 3\mathbf{i} - \mathbf{j}$

a $2\mathbf{a}$ **b** $\mathbf{a} + \mathbf{b}$

Step 1: Rewrite in vector notation and simplify.

a $2\mathbf{a} = 2(2\mathbf{i} + 4\mathbf{j})$
$= 4\mathbf{i} + 8\mathbf{j}$

b $\mathbf{a} + \mathbf{b} = (2\mathbf{i} + 4\mathbf{j}) + (3\mathbf{i} - \mathbf{j})$
$= 2\mathbf{i} + 3\mathbf{i} + 4\mathbf{j} - \mathbf{j}$
$= 5\mathbf{i} + 3\mathbf{j}$

The zero vector, 0

The **0** vector in **i**−**j** form is $0\mathbf{i} + 0\mathbf{j}$.

Equal vectors

If $a_1\mathbf{i} + a_2\mathbf{j} = b_1\mathbf{i} + b_2\mathbf{j}$ then $a_1 = b_1$ and $a_2 = b_2$.

The **i** components and the **j** components of each vector are equal.

Recall:
Equal vectors have the same size and direction.

Example 2.7 Given $\mathbf{a} = x\mathbf{i} + 3\mathbf{j}$ and $\mathbf{b} = 5\mathbf{i} - y\mathbf{j}$, find x and y if:

a $\mathbf{a} = 2\mathbf{b}$ **b** $2\mathbf{a} + \mathbf{b} = \mathbf{0}$

Step 1: Rewrite in vector notation and simplify.

a $x\mathbf{i} + 3\mathbf{j} = 2(5\mathbf{i} - y\mathbf{j})$
$x\mathbf{i} + 3\mathbf{j} = 10\mathbf{i} - 2y\mathbf{j}$

Step 2: Solve for unknowns.

i: $x = 10$
j: $3 = -2y$
$y = -\frac{3}{2}$

b $2(x\mathbf{i} + 3\mathbf{j}) + 5\mathbf{i} - y\mathbf{j} = 0\mathbf{i} + 0\mathbf{j}$
$(2x + 5)\mathbf{i} + (6 - y)\mathbf{j} = 0\mathbf{i} + 0\mathbf{j}$

i: $2x + 5 = 0$
$x = -\frac{5}{2}$

j: $6 - y = 0$
$y = 6$

Note:
Use the rules for addition and scalar multiplication.

Note:
Equate **i** and **j** components separately.

Parallel vectors

If $a_1\mathbf{i} + a_2\mathbf{j}$ is parallel to $b_1\mathbf{i} + b_2\mathbf{j}$, then $a_1\mathbf{i} + a_2\mathbf{j} = k(b_1\mathbf{i} + b_2\mathbf{j})$ where k is a constant.

Recall:
If vectors $\mathbf{u}$ and $\mathbf{v}$ are parallel, then $\mathbf{u} = k\mathbf{v}$ for some constant k.

Example 2.8 $\mathbf{u} = x\mathbf{i} + 2\mathbf{j}$ and $\mathbf{v} = 9\mathbf{i} - 6\mathbf{j}$. Find x if:

a $\mathbf{u}$ and $\mathbf{v}$ are parallel

b $\mathbf{u} + 3\mathbf{v}$ is parallel to the vector $\mathbf{i} + 3\mathbf{j}$.

Step 1: Rewrite in vector notation and simplify.

a
$$\mathbf{u} = k\mathbf{v}$$
$$x\mathbf{i} + 2\mathbf{j} = k(9\mathbf{i} - 6\mathbf{j})$$
$$x\mathbf{i} + 2\mathbf{j} = 9k\mathbf{i} - 6k\mathbf{j}$$

b
$$\mathbf{u} + 3\mathbf{v} = k(\mathbf{i} + 3\mathbf{j})$$
$$x\mathbf{i} + 2\mathbf{j} + 3(9\mathbf{i} - 6\mathbf{j}) = k(\mathbf{i} + 3\mathbf{j})$$
$$(x + 27)\mathbf{i} - 16\mathbf{j} = k\mathbf{i} + 3k\mathbf{j}$$

Note:
Write $\mathbf{u} = k\mathbf{v}$ for parallel vectors.

Step 2: Solve for unknowns.

$\mathbf{i}$: $x = 9k$

$\mathbf{j}$: $2 = -6k$

$k = -\frac{1}{3}$

$x = 9k$

$x = -3$

$\mathbf{i}$: $x + 27 = k$

$\mathbf{j}$: $-16 = 3k$

$k = -\frac{16}{3}$

$x = \frac{16}{3} - 27$

$x = -\frac{65}{3}$

Note:
Equate $\mathbf{i}$ and $\mathbf{j}$ components separately.

Unit vectors

Any vector with a magnitude of 1 is called a **unit vector**. For example, the vector $0.6\mathbf{i} + 0.8\mathbf{j}$ has magnitude $= \sqrt{0.6^2 + 0.8^2} = 1$ and so is a unit vector. You can find the unit vector in the direction of any vector $\mathbf{a}$ by dividing $\mathbf{a}$ by its magnitude, $|\mathbf{a}|$,

i.e. the unit vector in direction of $\mathbf{a}$, written $\hat{\mathbf{a}} = \dfrac{1}{|\mathbf{a}|}\mathbf{a}$.

Recall:
$|\mathbf{a}| = a$.

Example 2.9 Find the unit vector in the direction of $\mathbf{v} = 9\mathbf{i} - 6\mathbf{j}$.

Step 1: Divide by the size of the vector.

The unit vector in the direction of $\mathbf{v}$, $\hat{\mathbf{v}} = \dfrac{1}{|\mathbf{v}|}\mathbf{v}$

$$= \frac{(9\mathbf{i} - 6\mathbf{j})}{\sqrt{9^2 + (-6)^2}}$$

Step 2: Simplify.

$$= \frac{(9\mathbf{i} - 6\mathbf{j})}{\sqrt{117}}$$

$$= 0.83\mathbf{i} - 0.55\mathbf{j} \text{ (2 s.f.)}$$

Note:
The unit vector in the direction of $\mathbf{v}$ is $\mathbf{v}$ divided by its modulus. Check the answer:
$[0.83...^2 + (-0.55...)^2] = 1$

Defining a vector, given its magnitude and direction

A vector can be defined as a direction vector scaled by a constant

$$\mathbf{v} = k(a\mathbf{i} + b\mathbf{j}).$$

The magnitude of the vector is then

$$|\mathbf{v}| = \sqrt{(ka)^2 + (kb)^2}.$$

Note:
If two vectors have the same direction, they are parallel.

Example 2.10 Find a vector, **v**, of modulus 10 in the direction of the vector $4\mathbf{i} + 3\mathbf{j}$ in the form $x\mathbf{i} + y\mathbf{j}$.

Step 1: Rewrite in vector notation and simplify.

$$\mathbf{v} = k(4\mathbf{i} + 3\mathbf{j}) = 4k\mathbf{i} + 3k\mathbf{j}$$
$$|\mathbf{v}| = (4k)^2 + (3k)^2$$

Step 2: Solve for unknowns.

$$|\mathbf{v}| = 10$$
$$(4k)^2 + (3k)^2 = 10^2$$
$$16k^2 + 9k^2 = 100$$
$$25k^2 = 100$$
$$k^2 = 4$$
$$k = 2$$
$$\mathbf{v} = 4k\mathbf{i} + 3k\mathbf{j}$$
$$= 8\mathbf{i} + 6\mathbf{j}$$

Note:
Find the vector parallel to $4\mathbf{i} + 3\mathbf{j}$ with modulus 10.

Recall:
$\mathbf{v} = k\mathbf{u}$, where k is a constant.

Recall:
If $\mathbf{v} = a\mathbf{i} + b\mathbf{j}$
$|\mathbf{v}| = \sqrt{a^2 + b^2}$

Note:
The negative square root would give a vector in the opposite direction but with the same size.

SKILLS CHECK 2A: Vectors

1 A man walks x km from O on a bearing of $\theta°$ to a point A. He then walks a further distance of y km on a bearing of $\alpha°$ to a point B. Find the distance OB and the bearing of O from B when:

  a $\theta = 090$, $\alpha = 180$, $x = 4$ and $y = 3$ **b** $\theta = 024$, $\alpha = 160$, $x = 7$ and $y = 2$

  c $\theta = 290$, $\alpha = 045$, $x = 5$ and $y = 2$

2 The diagram shows a parallelogram $OPQR$. The vector **a** represents $\overrightarrow{RQ}$ and the vector **b** represents $\overrightarrow{PQ}$. M is the midpoint of OP and N is the midpoint of OR.

  Find, in terms of **a** and **b**, the following vectors:

  a $\overrightarrow{OQ}$ **b** $\overrightarrow{OP}$ **c** $\overrightarrow{OM}$

  d $\overrightarrow{MQ}$ **e** $\overrightarrow{QN}$ **f** $\overrightarrow{MN}$

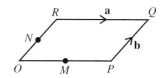

3 The diagram shows a triangle OAB with $\overrightarrow{OA} = \mathbf{a}$ and $\overrightarrow{OB} = \mathbf{b}$. The point C is on AB such that $\overrightarrow{AC} = 3\overrightarrow{CB}$.

  Express the following vectors in terms of **a** and **b**:

  a $\overrightarrow{AB}$ **b** $\overrightarrow{BA}$ **c** $\overrightarrow{AC}$ **d** $\overrightarrow{OC}$

4 Find the modulus and the angle that each of the following vectors makes with the vector **i**:

  a $3\mathbf{i} + 4\mathbf{j}$ **b** $6\mathbf{i} - \mathbf{j}$ **c** $-2\mathbf{i} - 2\mathbf{j}$ **d** $-3\mathbf{i} + 6\mathbf{j}$

 5 Find the following vectors in the form $x\mathbf{i} + y\mathbf{j}$ which have magnitude M and act at an angle $\theta°$ to the vector **i** where:

  a $M = 9$, $\theta = 20$ **b** $M = 4$, $\theta = 170$ **c** $M = 14$, $\theta = 210$

6 Given the vector $\mathbf{a} = 3\mathbf{i} + \mathbf{j}$, the vector $\mathbf{b} = 2\mathbf{i} - 4\mathbf{j}$ and the vector $\mathbf{c} = x\mathbf{i} + y\mathbf{j}$, find **c** in terms of **i** and **j**, the magnitude of **c** and the smaller angle that **c** makes with the vector **j** if:

  a $\mathbf{c} = \mathbf{a} + \mathbf{b}$ **b** $\mathbf{c} = \mathbf{a} - \mathbf{b}$ **c** $\mathbf{c} = 2\mathbf{a} + \mathbf{b}$

  d $\mathbf{c} = 2\mathbf{b} - 3\mathbf{a}$ **e** $\mathbf{c} + \mathbf{b} = 2\mathbf{a}$

 7 Find x if: **a** $2\mathbf{i} + 6\mathbf{j}$ is parallel to $x\mathbf{i} + 12\mathbf{j}$

  b $4\mathbf{i} - 3\mathbf{j}$ is parallel to $11\mathbf{i} + x\mathbf{j}$

  c $|x\mathbf{i} + 5\mathbf{j}| = 13$

8 Find the unit vector in the direction of: **a** $6\mathbf{i} + 8\mathbf{j}$ **b** $4\mathbf{i} - 8\mathbf{j}$

 9 Find the vector of modulus M in the direction of the vector $7\mathbf{i} + 24\mathbf{j}$ where:

      a $M = 1$ (unit vector)   **b** $M = 50$ **c** $M = 75$

10 Given that $\mathbf{a} = 2\mathbf{i} + \mathbf{j}$ and $\mathbf{b} = 3\mathbf{i} - 12\mathbf{j}$, find μ if:

      a $\mathbf{a} + \mu\mathbf{b}$ is parallel to the vector $3\mathbf{i} - 4\mathbf{j}$ **b** $\mu\mathbf{a} + \mathbf{b}$ is parallel to the vector $\mathbf{j}$.

SKILLS CHECK   **2A EXTRA is on the CD**

2.2 Application of vectors

Application of vectors to displacements, velocities and accelerations, and forces in a plane.

The displacement vector

The vectors used in Section 2.1 are **displacement vectors**; they describe relative positions. When measured from the origin the displacement vector is called a **position vector**.

Example 2.11   A particle is at point A (3, 4) in the $x-y$ plane. After two seconds the particle has moved to point B, where $\overrightarrow{AB} = 3\mathbf{i} - \mathbf{j}$. Find the position vector $\overrightarrow{OB}$.

Step 1: Use the vector formula and simplify.
$$\overrightarrow{OB} = \overrightarrow{OA} + \overrightarrow{AB}$$
$$= 3\mathbf{i} + 4\mathbf{j} + 3\mathbf{i} - \mathbf{j} = (6\mathbf{i} + 3\mathbf{j})$$

> **Note:**
> The coordinates (3, 4) represent the position vector ($3\mathbf{i} + 4\mathbf{j}$).

Relative displacement

If the point A is described by the position vector $\mathbf{r}_A$ and the point B by the position vector $\mathbf{r}_B$, then the position vector of A relative to B is:
$$\overrightarrow{BA} = \mathbf{r}_A - \mathbf{r}_B$$
This is the displacement vector from B to A. The distance between A and B is $|\mathbf{r}_A - \mathbf{r}_B|$.

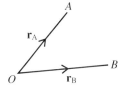

> **Recall:**
> $\overrightarrow{BA} = -\overrightarrow{OB} + \overrightarrow{OA}$

Example 2.12   At a given time the position vectors of particles A and B relative to a fixed origin O are $(2\mathbf{i} + 6\mathbf{j})$ m and $(3\mathbf{i} - 3\mathbf{j})$ m. Find:

   a   the position vector of A relative to B

   b   the distance between them at this time.

Step 1: Use the vector formula.

Step 2: Simplify.

   a $\overrightarrow{BA} = \mathbf{r}_A - \mathbf{r}_B$ **b** $|\overrightarrow{BA}| = |\mathbf{r}_A - \mathbf{r}_B|$

        $= 2\mathbf{i} + 6\mathbf{j} - (3\mathbf{i} - 3\mathbf{j})$ $= \sqrt{(-1)^2 + 9^2}$

        $= -\mathbf{i} + 9\mathbf{j}$ $= 9.06$ m (3 s.f.)

The velocity vector

Velocity is a vector quantity; it is the rate of change of displacement with time. The magnitude of velocity is called the **speed**.
When the velocity is constant (no acceleration):

$$\text{velocity} = \frac{\text{change in displacement}}{\text{time taken}}$$

A particle moving $(x\mathbf{i} + y\mathbf{j})$ m every second has a **velocity** $(x\mathbf{i} + y\mathbf{j})$ m/s.

> **Note:**
> This is the vector form of the scalar relationship:
> $$\text{speed} = \frac{\text{distance}}{\text{time}}$$

> **Note:**
> You can also write m/s as m s^{-1}.

Example 2.13 At time $t = 0$ s, a particle is at a position $(6\mathbf{i} - 7\mathbf{j})$ m relative to the origin O. At $t = 3$ s the particle is at a position $(3\mathbf{i} + 2\mathbf{j})$ m relative to the origin O. Given that the velocity is constant find:

a the velocity   **b** the speed of the particle.

Step 1: Use the vector formula.

Step 2: Simplify.

a $\mathbf{v} = \dfrac{(3\mathbf{i} + 2\mathbf{j}) - (6\mathbf{i} - 7\mathbf{j})}{3}$

$= \dfrac{-3\mathbf{i} + 9\mathbf{j}}{3}$ **b** $|\mathbf{v}| = |-\mathbf{i} + 3\mathbf{j}| = \sqrt{(-1)^2 + 3^2}$

$= (-\mathbf{i} + 3\mathbf{j})$ $= 3.16 \ (3 \text{ s.f.})$

The velocity is $(-\mathbf{i} + 3\mathbf{j})$ m s^{-1}.   The speed is 3.16 m s^{-1}.

Tip:
Don't forget the units.

The acceleration vector

The acceleration of a particle is the rate of change of the velocity with time. When the acceleration is constant:

$$\text{acceleration} = \frac{\text{change in velocity}}{\text{time taken}}$$

The units for acceleration are m/s^2 or m s^{-2}.

Example 2.14 A particle is initially travelling with velocity $(-2\mathbf{i} - 9\mathbf{j})$ m s^{-1} and 2 seconds later it has velocity $(6\mathbf{i} - 11\mathbf{j})$ m s^{-1}, where $\mathbf{i}$ and $\mathbf{j}$ are unit vectors in the directions of the positive x- and y-axes, respectively. Given that the acceleration of the particle is constant find:

a the acceleration

b the magnitude of the acceleration

c the angle that the acceleration makes with the vector $\mathbf{j}$.

Step 1: Use the vector formula.

Step 2: Simplify.

a $\mathbf{a} = \dfrac{(6\mathbf{i} - 11\mathbf{j}) - (-2\mathbf{i} - 9\mathbf{j})}{2}$ **b** $|\mathbf{a}| = |4\mathbf{i} - \mathbf{j}| = \sqrt{4^2 + (-1)^2}$

$= \dfrac{8\mathbf{i} - 2\mathbf{j}}{2}$ $= \sqrt{17}$

Step 1: Draw a clear diagram, marking all known angles and distances.

$= (4\mathbf{i} - \mathbf{j})$

The acceleration is $(4\mathbf{i} - \mathbf{j})$ m s^{-2}.

The magnitude of the acceleration is $\sqrt{17}$ m s^{-2}.

Step 2: Use trigonometry to calculate unknown angles.

c $\tan \theta° = \frac{1}{4}$

$\theta = 14$ (nearest degree)

The angle made with the vector $\mathbf{j}$ is $(90° + 14°) = 104°$.

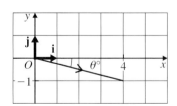

Constant velocity

A particle at point A with position vector $\mathbf{r_0}$ travels with constant velocity $\mathbf{v}$ m s^{-1}. In t seconds it has moved a displacement of $(\mathbf{v}t)$ m from point A to the point B. This can be drawn as a vector diagram.

Note:
The velocity must be constant and so the acceleration is 0 m s^{-2}.

The position vector $\mathbf{r}$ can be calculated as:

$\mathbf{r} = \mathbf{r_0} + t\mathbf{v}$.

This equation calculates the position vector after t seconds of a particle travelling with constant velocity.

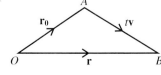

Recall:
Displacement = velocity $\times$ time

Note:
The units must be consistent.

Example 2.15 At time $t = 0$, where t is time (in seconds), a particle is at the point with position vector $(4\mathbf{i} - \mathbf{j})$ m and travels with velocity $(-2\mathbf{i} + 2\mathbf{j})$ m s^{-1}. Find:

a the position vector of the particle after t seconds

b the distance the particle is from O after 3 seconds.

Step 1: Summarise the information in a vector diagram. Identify $\mathbf{r_0}$ and $\mathbf{v}$.

$\mathbf{r_0} = (4\mathbf{i} - \mathbf{j})$

$\mathbf{v} = (-2\mathbf{i} + 2\mathbf{j})$

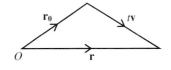

Step 2: Rewrite in vector notation. Use $\mathbf{r} = \mathbf{r_0} + t\mathbf{v}$.

a $\mathbf{r} = \mathbf{r_0} + t\mathbf{v}$

$= (4\mathbf{i} - \mathbf{j}) + t(-2\mathbf{i} + 2\mathbf{j})$

$= [(4 - 2t)\mathbf{i} + (2t - 1)\mathbf{j}]$

b At $t = 3$

$\mathbf{r} = (-2\mathbf{i} + 5\mathbf{j})$

$|-2\mathbf{i} + 5\mathbf{j}| = \sqrt{(-2)^2 + 5^2}$

Tip
Tip:

Combine the $\mathbf{i}$ and $\mathbf{j}$ components.

Step 3: Solve for unknowns.

$= \sqrt{29} = 5.4$ (2 s.f.)

The position vector of the particle is $[(4 - 2t)\mathbf{i} + (2t - 1)\mathbf{j}]$ m.   The distance of the particle from O is 5.4 m.

Sometimes the information may not be given in $\mathbf{i} - \mathbf{j}$ form and you may be required to find the position vector of the particle after time t.

Example 2.16 A particle with a position vector $(3\mathbf{i} - \mathbf{j})$ m travels with speed 26 m s^{-1} in the direction of the vector $(5\mathbf{i} - 12\mathbf{j})$, where $\mathbf{i}$ and $\mathbf{j}$ are unit vectors due east and north respectively. Find the position vector of the particle after t seconds, in the form $\mathbf{r} = \mathbf{r_0} + t\mathbf{v}$.

Step 1: Summarise the information in a vector diagram. Identify $\mathbf{r_0}$ and $\mathbf{v}$.

$\mathbf{r_0} = (3\mathbf{i} - \mathbf{j})$

$\mathbf{v} = k(5\mathbf{i} - 12\mathbf{j})$

$|\mathbf{v}| = 26$

Recall:

Example 2.10 to find the velocity, given its speed and direction.

Step 2: Rewrite in vector notation. Use $\mathbf{r} = \mathbf{r_0} + t\mathbf{v}$.

$|\mathbf{v}| = 26$

$(5k)^2 + (-12k)^2 = 26^2$

Step 3: Solve for unknowns.

$k = 2$

$\mathbf{v} = (10\mathbf{i} - 24\mathbf{j})$

$\mathbf{r} = \mathbf{r_0} + t\mathbf{v}$

$= [(3\mathbf{i} - \mathbf{j}) + t(10\mathbf{i} - 24\mathbf{j})]$

Recall:

Example 2.6; Example 2.10.

The position vector is $[(3\mathbf{i} - \mathbf{j}) + t(10\mathbf{i} - 24\mathbf{j})]$ m.

Example 2.17 A particle starts at a point 8 m from O at an angle of $45°$ anticlockwise from east and travels with velocity $(-2\mathbf{i} - 3\mathbf{j})$ m s^{-1}, where $\mathbf{i}$ and $\mathbf{j}$ are unit vectors due east and north respectively. Find the position vector of the particle after t seconds in the form $\mathbf{r} = \mathbf{r_0} + t\mathbf{v}$.

Step 1: Summarise the information in a vector diagram. Identify $\mathbf{r_0}$ and $\mathbf{v}$.

$\mathbf{r_0} = M \cos \theta° \mathbf{i} + M \sin \theta° \mathbf{j}$,

where $M = 8$ and $\theta = 45$

Step 2: Rewrite in vector notation and simplify. Use $\mathbf{r} = \mathbf{r_0} + t\mathbf{v}$.

$\mathbf{v} = (-2\mathbf{i} - 3\mathbf{j})$

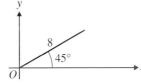

Recall:

From C2, $\cos 45° = \sin 45°$

$= \dfrac{1}{\sqrt{2}} = \dfrac{\sqrt{2}}{2}$.

So $8 \cos 45° = 8 \times \dfrac{\sqrt{2}}{2} = 4\sqrt{2}$.

Step 3: Solve for unknowns.

$\mathbf{r_0} = 8 \cos 45° \, \mathbf{i} + 8 \sin 45° \, \mathbf{j}$

$= (4\sqrt{2}\mathbf{i} + 4\sqrt{2}\mathbf{j})$

$\mathbf{r} = \mathbf{r_0} + t\mathbf{v}$

$= [(4\sqrt{2}\mathbf{i} + 4\sqrt{2}\mathbf{j}) + t(-2\mathbf{i} - 3\mathbf{j})]$

The position vector is $[(4\sqrt{2}\mathbf{i} + 4\sqrt{2}\mathbf{j}) + t(-2\mathbf{i} - 3\mathbf{j})]$ m.

Example 2.18 At noon, John is initially at point O and walks with constant velocity $(\mathbf{i} + 3\mathbf{j})$ km h^{-1}, where $\mathbf{i}$ and $\mathbf{j}$ are unit vectors due east and due north respectively. At the same time Karim, who is initially at the point with position vector $(14\mathbf{i} + 2\mathbf{j})$ km, is walking with constant velocity $(-3\mathbf{i} - 9\mathbf{j})$ km h^{-1}. Find:

a the position vector of each person after t hours

b the position vector of each person at 1.30 pm

c the position vector of Karim relative to John after t hours.

d Given that the distance between John and Karim t hours after noon is d km, show that $d^2 = 160t^2 - 160t + 200$.

e Find the least distance between John and Karim and the time at which this occurs.

Note:
Questions may involve use of km h^{-1} instead of ms^{-1}. Be consistent within the question.

Step 1: Summarise the information in a vector diagram. Identify $\mathbf{r_0}$ and $\mathbf{v}$.

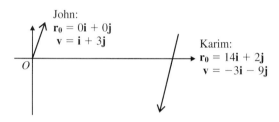

John:
$\mathbf{r_0} = 0\mathbf{i} + 0\mathbf{j}$
$\mathbf{v} = \mathbf{i} + 3\mathbf{j}$

Karim:
$\mathbf{r_0} = 14\mathbf{i} + 2\mathbf{j}$
$\mathbf{v} = -3\mathbf{i} - 9\mathbf{j}$

Step 2: Rewrite in vector notation, simplify. Use $\mathbf{r} = \mathbf{r_0} + t\mathbf{v}$.

a

For John,

$\mathbf{r_J} = \mathbf{r_0} + t\mathbf{v}$
$= (0\mathbf{i} + 0\mathbf{j}) + t(\mathbf{i} + 3\mathbf{j})$
$= [t\mathbf{i} + 3t\mathbf{j}]$

For Karim,

$\mathbf{r_K} = \mathbf{r_0} + t\mathbf{v}$
$= (14\mathbf{i} + 2\mathbf{j}) + t(-3\mathbf{i} - 9\mathbf{j})$
$= [(14 - 3t)\mathbf{i} + (2 - 9t)\mathbf{j}]$

The position vectors for John and Karim are $[t\mathbf{i} + 3t\mathbf{j}]$ km and $[(14 - 3t)\mathbf{i} + (2 - 9t)\mathbf{j}]$ km respectively.

Note:
Let $\mathbf{r_J}$ be John's position vector and $\mathbf{r_K}$ be Karim's position vector.

Step 3: Solve for unknowns by substituting for t.

b At 1.30pm, $t = 1.5$ hours:

$\mathbf{r_J} = (1.5\mathbf{i} + 4.5\mathbf{j})$ $\qquad$ $\mathbf{r_K} = (9.5\mathbf{i} - 11.5\mathbf{j})$

Step 4: Use the fact that the position vector of A relative to $B = \mathbf{r_A} - \mathbf{r_B}$.

c $\mathbf{r_K} - \mathbf{r_J} = [(14 - 3t)\mathbf{i} + (2 - 9t)\mathbf{j}] - [t\mathbf{i} + 3t\mathbf{j}]$
$= [(14 - 4t)\mathbf{i} + (2 - 12t)\mathbf{j}]$

The position vector of Karim relative to John after t hours is $[(14 - 4t)\mathbf{i} + (2 - 12t)\mathbf{j}]$ km.

Recall:
Example 2.7, add/subtract the $\mathbf{i}$ components and the $\mathbf{j}$ components separately.

Step 5: Use distance between A and B $= |\mathbf{r_A} - \mathbf{r_B}|$.

d $d^2 = |\mathbf{r_K} - \mathbf{r_J}|^2 = (14 - 4t)^2 + (2 - 12t)^2$
$= 196 - 112t + 16t^2 + 4 - 48t + 144t^2$
$= 160t^2 - 160t + 200$

e Let $y = d^2$, then $y = 160t^2 - 160t + 200$

$$\frac{dy}{dt} = 320t - 160$$

$$0 = 320t - 160$$

$$t = 0.5$$

This is the time when Karim and John are closest together.

Substitute $t = 0.5$ into $d^2 = 160t^2 - 160t + 200$

$$= 160$$

$$d = 12.6\ldots$$

So, the time when they are closest together is 12.30 pm and the closest distance between them is 12.6 km (3 s.f.).

Recall:
Differentiation from C2: to find maxima or minima, set $\frac{dy}{dt} = 0$ and solve for t.

Note:
You can also use completing the square to find the minimum of a quadratic if you have not yet covered differentiation in C2.

If two particles collide then the position vectors of each particle are equal.

Example 2.19   At $t = 0$, where t is the time measured in hours, a helicopter leaves from a point with position vector $(2\mathbf{i} + 4\mathbf{j})$ km and travels with constant velocity $(4\mathbf{i} - \mathbf{j})$ km h^{-1} where $\mathbf{i}$ and $\mathbf{j}$ are unit vectors, 1 km due east and 1 km due north respectively.

   a   Find the position vector of the helicopter after t hours.

At $t = 0$ another helicopter leaves from another point with position vector $(-4\mathbf{i})$ km and travels with constant velocity $(7\mathbf{i} + \mathbf{j})$ km h^{-1}.

   b   Find the position vector of the second helicopter after t hours.

   c   Find whether the helicopters collide, and if they do collide, find the position vector of their point of intersection.

Step 1: Summarise the information in a vector diagram. Identify $\mathbf{r_0}$ and $\mathbf{v}$.

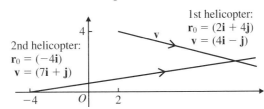

Step 2: Rewrite in vector notation and simplify. Use $\mathbf{r} = \mathbf{r_0} + t\mathbf{v}$.

a   For the first helicopter:

$$\mathbf{r_1} = \mathbf{r_0} + t\mathbf{v}$$
$$= (2\mathbf{i} + 4\mathbf{j}) + t(4\mathbf{i} - \mathbf{j})$$
$$= [(2 + 4t)\mathbf{i} + (4 - t)\mathbf{j}]$$

The position vector is $[(2 + 4t)\mathbf{i} + (4 - t)\mathbf{j}]$ km.

b   For the second helicopter:

$$\mathbf{r_2} = \mathbf{r_0} + t\mathbf{v}$$
$$= (-4\mathbf{i}) + t(7\mathbf{i} + \mathbf{j})$$
$$= [(7t - 4)\mathbf{i} + t\mathbf{j}]$$

The position vector is $= [(7t - 4)\mathbf{i} + t\mathbf{j}]$ km.

> **Recall:**
> Example 2.7 for equal vectors.

Step 3: Solve for unknowns by equating $\mathbf{r_1}$ and $\mathbf{r_2}$ for collisions.

c   If the helicopters collide then: $\mathbf{r_1} = \mathbf{r_2}$

$$(2 + 4t)\mathbf{i} + (4 - t)\mathbf{j} = (7t - 4)\mathbf{i} + t\mathbf{j}$$

Equating components:

$\mathbf{i}$: $2 + 4t = 7t - 4$

$\qquad t = 2$

$\mathbf{j}$: $\quad 4 - t = t$

$\qquad t = 2$

> **Note:**
> If the times for each component are different then they do not collide.

The time at which the $\mathbf{i}$ components are equal is the same time at which the $\mathbf{j}$ components are equal and so the helicopters do collide.

Substitute $t = 2$ into $\mathbf{r_1} = [(2 + 4t)\mathbf{i} + (4 - t)\mathbf{j}]$

$$\mathbf{r_1} = (10\mathbf{i} + 2\mathbf{j})$$

The helicopters collide at $t = 2$ hours at the point $(10\mathbf{i} + 2\mathbf{j})$ km.

> **Tip:**
> To check your answer, substitute $t = 2$ into $\mathbf{r_2}$ to give the same point.

SKILLS CHECK **2B: Application of vectors**

1   Find the constant velocity and the speed of a particle that moves from position vectors $\mathbf{r_A}$ m to $\mathbf{r_B}$ m in t seconds where:

   a $\mathbf{r_A} = 5\mathbf{i} + \mathbf{j}$, $\mathbf{r_B} = 7\mathbf{i} + 3\mathbf{j}$, $t = 2$

   b $\mathbf{r_A} = 5\mathbf{i} - \mathbf{j}$, $\mathbf{r_B} = 2\mathbf{i} + 5\mathbf{j}$, $t = 3$

   c $\mathbf{r_A} = 4\mathbf{i} - 6\mathbf{j}$, $\mathbf{r_B} = 13\mathbf{i} + 24\mathbf{j}$, $t = 6$.

2 A particle travels with constant velocity $(7\mathbf{i} + 8\mathbf{j})$ m s^{-1} starting from the point with position vector $(3\mathbf{i} + \mathbf{j})$ m relative to a fixed origin O. Find:

   a the displacement of the particle after 3 seconds

   b the position vector of the particle after 3 seconds

   c the distance that the particle is from O after 3 seconds and the angle that the position vector makes with the vector $\mathbf{i}$.

3 A car travels with uniform acceleration. Initially the car has velocity $\mathbf{v}_1$ m s^{-1} and after t seconds it has velocity $\mathbf{v}_2$ m s^{-1}. Find the acceleration vector of the particle where:

   a $\mathbf{v}_1 = (6\mathbf{i} + 3\mathbf{j})$, $\mathbf{v}_2 = (9\mathbf{i} + 12\mathbf{j})$, $t = 2$

   b $\mathbf{v}_1 = (-77\mathbf{i} + 32\mathbf{j})$, $\mathbf{v}_2 = (-13\mathbf{i} - 32\mathbf{j})$, $t = 4$

   c $\mathbf{v}_1 = (-\mathbf{i} + \mathbf{j})$, $\mathbf{v}_2 = (-\mathbf{i} + 6\mathbf{j})$, $t = 5$.

 4 A ball is rolling across a smooth table. It accelerates uniformly at $(2\mathbf{i} + \mathbf{j})$ m s^{-2}, where $\mathbf{i}$ and $\mathbf{j}$ are unit vectors due east and due north respectively. The final velocity of the ball after travelling with this acceleration for 4 seconds is $(-4\mathbf{i} - 12\mathbf{j})$ m s^{-1}. Find the initial speed of the ball.

5 At $t = 0$, where t is the time measured in seconds, a particle sets out from a point A with position vector $(3\mathbf{i} + 8\mathbf{j})$ m and travels with velocity $(\mathbf{i} - 2\mathbf{j})$ m s^{-1}.

   a Find the position vector of the particle after t seconds.

After three seconds the particle reaches the point B.

   b Find the position vector of the point B.

   c Show that the distance of OB, where O is the origin, can be written as $a\sqrt{10}$ m. State the value of a.

6 a Find the displacement of A relative to B if the position vector of the point A is $(9\mathbf{i} + 11\mathbf{j})$ m and the position vector of the point B is $(6\mathbf{i} - 3\mathbf{j})$ m. Hence, find the distance AB.

   b Particle A is travelling with velocity $\mathbf{v}_A = (4\mathbf{i} + 7\mathbf{j})$, measured in m s^{-1}, while particle B has speed 10 m s^{-1} in the direction of the vector $(3\mathbf{i} + 4\mathbf{j})$. Find $\mathbf{v}_B$, the velocity of particle B, and hence find the vector $(\mathbf{v}_A - \mathbf{v}_B)$.

   c Find the position vector of A if the position vector of A relative to B is $(5\mathbf{i} + 3\mathbf{j})$ m, and the position vector of B is $(2\mathbf{i} + \mathbf{j})$ m.

 7 In this question $\mathbf{i}$ and $\mathbf{j}$ are unit vectors due east and due north, respectively.

At noon, a ship P leaves a port and sails with constant velocity $(\mathbf{i} + 3\mathbf{j})$ km h^{-1}.

   a By taking the port as the origin, write down the position vector of the ship after t hours.

At noon another ship Q is at the point with position vector $(10\mathbf{i} + 10\mathbf{j})$ km and travels with constant velocity $(-\mathbf{i} + 2\mathbf{j})$ km h^{-1}.

   b Find the position vector of ship Q after t hours.

   c Find the distance between the two ships at 14:00 hours.

   d Show that the distance, d km, between the two ships after t hours is given by:

$$d^2 = 5t^2 - 60t + 200$$

   e By finding the minimum value of d^2, find the time at which the distance between the ships is least and find the closest distance between the two ships.

8 A cricket ball is hit from a point with position vector $(7\mathbf{i} - 6\mathbf{j})$ m relative to a fixed origin O, where $\mathbf{i}$ and $\mathbf{j}$ are unit vectors due east and north respectively. The ball travels along a smooth ground with constant velocity $(-11\mathbf{i} + 4\mathbf{j})$ m s^{-1}. A fielder is standing at the point with position vector $(2\mathbf{i} - 3\mathbf{j})$ m and attempts to intercept the cricket ball. He leaves at the same time as the ball is hit. The fielder runs with speed $\sqrt{37}$ m s^{-1} in the direction of the vector $(-12\mathbf{i} + 2\mathbf{j})$. Find whether the cricketer will intercept the ball and, if he does, find the position vector of the point of interception. Comment on any assumptions that have been made in modelling this situation.

 9 In this question $\mathbf{i}$ and $\mathbf{j}$ are unit vectors and are at right angles to each other.

A cyclist leaves a point with position vector $(a\mathbf{i} + b\mathbf{j})$ m and travels with velocity $(2\mathbf{i} + \mathbf{j})$ m s^{-1}. A motorcyclist leaves at the same time from a point with position vector $(\mathbf{i} + 13\mathbf{j})$ m and travels with velocity $(7\mathbf{i} - 3\mathbf{j})$ m s^{-1}. Given that the cyclist and the motorcyclist collide after three seconds, find the values of the constants a and b.

In order to avoid a collision, after two seconds the cyclist decides to change his velocity to $(2\mathbf{i} + 2\mathbf{j})$ m s^{-1}. Using the values of a and b found above, find the distance between the particles after 3 seconds.

SKILLS CHECK **2B EXTRA is on the CD**

Examination practice Vectors in mechanics

1 A particle P is moving with constant velocity $(5\mathbf{i} - 3\mathbf{j})$ m s^{-1}. At time $t = 0$, its position vector, with respect to a fixed origin O, is $(-2\mathbf{i} + \mathbf{j})$ m. Find, to 3 significant figures,

a the speed of P,

b the distance of P from O when $t = 2$ s. [Edexcel Jan 2000]

 2 A particle P moves in a straight line with constant velocity. Initially P is at the point A with position vector $(2\mathbf{i} - \mathbf{j})$ m relative to a fixed origin O, and 2 s later it is at the point B with position vector $(6\mathbf{i} + \mathbf{j})$ m.

a Find the velocity of P.

b Find, in degrees to one decimal place, the size of the angle between the direction of motion of P and the vector $\mathbf{i}$.

Three seconds after it passes B the particle P reaches the point C.

c Find, in m to one decimal place, the distance OC. [Edexcel Jan 2001]

3 A destroyer is moving due west at a constant speed of 10 km h^{-1}. It has radar on board which, at time $t = 0$, identifies a cruiser, 50 km due west and moving due north with a constant speed of 20 km h^{-1}. The unit vectors $\mathbf{i}$ and $\mathbf{j}$ are directed due east and north respectively, and the origin O is taken to be the initial position of the destroyer. Each vessel maintains its constant velocity.

a Write down the velocity of each vessel in vector form.

b Find the position of each vessel at time t hours.

c Show that the distance d km between the vessels at time t hours is given by
$$d^2 = 500t^2 - 1000t + 2500.$$

The radar on the cruiser detects vessels up to a distance of 40 km. By finding the minimum value of d^2, or otherwise,

d determine whether the destroyer will be detected by the cruiser's radar. [Edexcel Jan 1998]

4 Two dogs, Albert and Bernard, are running in a field. The field is a horizontal plane and O is a fixed point on the field from which all position vectors are measured. The perpendicular vectors $\mathbf{i}$ and $\mathbf{j}$ are unit vectors in the plane. At time $t = 0$, Albert is at the point O and Bernard is at the point with position vector $2\mathbf{j}$ m. Albert runs with constant velocity towards the point with position vector $(24\mathbf{i} + 18\mathbf{j})$ m, and Bernard runs, also with constant velocity, towards the point with position vector $(-8\mathbf{i} + 26\mathbf{j})$ m. Albert's speed is 10 m s^{-1}, and Bernard's speed is $3\sqrt{10}$ m s^{-1}.

  a Find the velocity of Albert, giving your answer in vector form.

  b Show that the velocity of Bernard is $(-3\mathbf{i} + 9\mathbf{j})$ m s^{-1}.

Given that the velocities of the two dogs remain constant,

  c find, in m to one decimal place, the distance between the dogs 2 s after they start.

[Edexcel Jan 1996]

5 Two motor boats A and B are moving with constant velocities. The velocity of A is 30 km h^{-1} due north, and B is moving at 20 km h^{-1} on a bearing of $60°$. The unit vectors $\mathbf{i}$ and $\mathbf{j}$ are directed due east and north respectively. At 10 a.m. the position vector of B is $70\mathbf{j}$ km relative to a fixed origin O and A is at the point O, t hours later, the position vectors of A and B are $\mathbf{r}$ km and $\mathbf{s}$ km respectively.

  a Find the velocity of B, in the form $(p\mathbf{i} + q\mathbf{j})$ km h^{-1}.

  b Find expressions for $\mathbf{r}$ and $\mathbf{s}$ in terms of t.

The boats can maintain radio contact with each other, provided that the distance between them is no more than 70 km.

  c Find the time at which the boats are again at the maximum distance at which they can maintain radio contact with each other.

[Edexcel June 1996]

3 Kinematics of a particle moving in a straight line

3.1 Motion in a horizontal plane

Motion in a straight line with constant acceleration.

The aim of kinematics is to analyse the motion of a particle that is travelling in a straight line – its velocity, time of travel, acceleration and the path that it follows. In this unit, the particle always has a constant acceleration.

The properties that you are analysing are (with standard units in brackets):

s = displacement of particle from a fixed point (metres, m)
u = the initial velocity of the particle (m/s or m s^{-1})
v = the final velocity of the particle (m/s or m s^{-1})
a = the acceleration of the particle (m/s^2 or m s^{-2})
t = time taken for the particle to travel (seconds, s).

The **equations of motion** contain the variables that you are aiming to study. Sometimes these are called the **uniform acceleration equations.** These are:

$$v = u + at$$
$$v^2 = u^2 + 2as$$
$$s = \frac{(u + v)}{2}t$$
$$s = ut + \tfrac{1}{2}at^2$$

Each equation contains four variables. In most questions you will be given three variables and asked to calculate the fourth.

Recall:
Acceleration is the rate of change of velocity of a particle, where velocity is the speed of a particle in a given direction.

Note:
These variables are often referred to as *suvat*.

Tip:
These are not in the formulae book; learn them.

Note:
The units must be consistent when using these equations.

Example 3.1 A particle moves in a straight line from A to B. The particle starts from rest at A and accelerates at 2 m s^{-2} until it reaches a speed of 8 m s^{-1} at B.

a Find how long it takes to travel from A to B.

b Find the distance AB.

Tip:
If a particle starts from rest, this is another way of saying $u = 0$.

Step 1: Draw a clear diagram to represent the information given.

```
0 m s⁻¹                    2 m s⁻²                      8 m s⁻¹
 A ────────────────────────→→───────────────────────── B
```

Note:
Acceleration is represented by a double arrow →→ and velocity by a single arrow →.

Step 2: Fill the information in *suvat*, identifying what is required with a question mark.

a
s = not required
$u = 0$
$v = 8$
$a = 2$
$t = ?$

b
$s = ?$
$u = 0$
$v = 8$
$a = 2$
t = not required

Step 3: Pick an equation of motion relating the three known variables with the unknown that is required, insert values, rearrange (if necessary) and solve.

a
$v = u + at$
$8 = 0 + 2t$
$t = 4$

It takes 4 s to travel from A to B.

b
$v^2 = u^2 + 2as$
$8^2 = 0^2 + 2(2)s$
$s = 16$

Distance AB is 16 m.

Tip:
Don't forget the units.

A particle that slows down is **decelerating**, in which case the acceleration is negative. If the acceleration of a particle is -7 m s^{-2}, then its **deceleration** or **retardation** is 7 m s^{-2}.

Example 3.2 A train is travelling at 60 km h^{-1} on a straight railway track. The driver of the train sees a red signal 500 m ahead. He immediately applies the brakes so that the train decelerates at 0.25 m s^{-2}.

a Find how far the train is from the signal when it comes to a halt.

b What is the velocity of the train as it passes the signal?

Step 1: Draw a clear diagram to represent the information given.

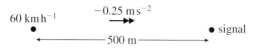

Step 2: Fill the information in *suvat*, identifying what is required with a question mark.

Note:
You must work in consistent units.
1000 m = 1 km and
3600 s = 1 hour, so to convert from km h^{-1} to m s^{-1} you multiply by 1000 and divide by 3600. In this case,
60 km h^{-1} = $16\frac{2}{3}$ m s^{-1}

a Motion of train from the start until it stops:

$s = ?$
$u = 16\frac{2}{3}$
$v = 0$
$a = -0.25$
t = not required

Step 3: Pick an equation of motion relating the three known variables with the unknown that is required, insert values, rearrange (if necessary) and solve.

$v^2 = u^2 + 2as$
$(0)^2 = (16\frac{2}{3})^2 + 2(-0.25)s$
$s = 556$ (3 s.f.)

The train stops $(556 - 500)$ m $= 56$ m after the signal.

b Motion of train from the start until it passes the signal:

$s = 500$
$u = 16\frac{2}{3}$
$v = ?$
$a = -0.25$
t = not required

$v^2 = u^2 + 2as$
$v^2 = (16\frac{2}{3})^2 + 2(-0.25)500$
$v = 5.27$ (3 s.f.)

The train travels at 5.27 m s^{-1} as it passes the red signal.

Note:
Be careful not to confuse the variable s (displacement) with the unit s (seconds).

Note:
Only the positive root is physically reasonable.

Example 3.3 Two cars, *A* and *B*, start from the same point *O* and travel in a straight line. Car *A* leaves at $t = 0$ seconds and travels with constant speed 5 m s^{-1}. Car *B* leaves 10 seconds later and accelerates uniformly at 5 m s^{-2}. Find the distance from *O* when car *B* overtakes car *A*.

Note:
A particle travels with constant speed is another way of saying $a = 0$.

Suppose they meet *T* seconds after car *A* leaves *O*. t_A and t_B are the times for each car, measured from the start of *A*'s journey:

Step 1: Draw a clear diagram to represent the information given.

For A:
$t_A = 0$
$u = 5$ $a = 0$ $t_A = T$

For B:
$t_B = 10$
$u = 0$ $a = 5$ $t_B = T$

Note:
Consider the journey to the point where both cars have travelled the same distance.

Step 2: Fill the information in *suvat*, identifying what is required with a question mark.

For car *A*:
$s = ?$
$u = 5$
v = not required
$a = 0$
$t = T$
$s = ut + \frac{1}{2}at^2$
$s = 5T + \frac{1}{2}(0)T^2$
$= 5T$

For car *B*:
$s = ?$
$u = 0$
v = not required
$a = 5$
$t = T - 10$
$s = ut + \frac{1}{2}at^2$
$s = (0)T + \frac{1}{2}(5)(T-10)^2$
$= 2.5T^2 - 50T + 250$

Note:
When there are two motions, apply *suvat* to each situation and find the common link. In this case, if car *A* has travelled for *T* s then car *B* has travelled for $(T - 10)$ s.

Note:
Pick equations that include both *s* and *t*.

Step 3: Pick an equation of motion relating the three known variables with the unknown that is required, insert values, rearrange (if necessary) and solve.

When *B* overtakes *A*
$5T = 2.5T^2 - 50T + 250$
$0 = 2.5T^2 - 55T + 250$
$T = 15.58...$ or $6.42...$

Recall:
Using the quadratic formula from C1.

$T = 15.58...$ is the time that they meet because $T = 6.42...$ would give a negative time *t* for car *B*.
The cars have travelled $5 \times 15.58... = 77.9$ m when they meet.

Example 3.4 A particle moves in a straight line. It passes a point O with velocity u m s^{-1} and has constant acceleration a m s^{-2}. Two seconds later it passes a point P. One second after it passes the point P it passes a point Q. Given that the distance OP is 34 m and the distance PQ is 20 m, find u and a.

Step 1: Draw a clear diagram to represent the information given.

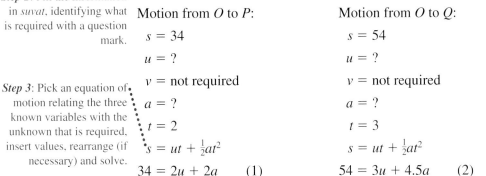

Step 2: Fill the information in *suvat*, identifying what is required with a question mark.

Motion from O to P:

$s = 34$

$u = ?$

$v =$ not required

$a = ?$

$t = 2$

Step 3: Pick an equation of motion relating the three known variables with the unknown that is required, insert values, rearrange (if necessary) and solve.

$s = ut + \frac{1}{2}at^2$

$34 = 2u + 2a \qquad (1)$

Motion from O to Q:

$s = 54$

$u = ?$

$v =$ not required

$a = ?$

$t = 3$

$s = ut + \frac{1}{2}at^2$

$54 = 3u + 4.5a \qquad (2)$

Solving (1) and (2) simultaneously gives $u = 15$ and $a = 2$.

Note:
Analyse the motion from O to P and from O to Q because both sets of equations will have the same two unknowns, u and a.

Note:
In this case there are two equations and two unknowns: solve them simultaneously.

SKILLS CHECK **3A: Motion in a horizontal plane**

1 A particle travels in a straight horizontal line. It accelerates uniformly from rest to a speed of 4 m s^{-1} in 2 seconds.

a Find the acceleration.

b Find the distance travelled by the particle.

2 An object undergoing uniform acceleration travels 100 m in 12 seconds. Given the initial velocity is zero, find the acceleration.

3 A particle moves along a straight line AB with constant acceleration of 1 m s^{-2}. If AB is 20 m in length and it takes 2 seconds to travel from A to B, what was the velocity of the particle at A?

 4 A car travels in a straight line with uniform acceleration of 3 m s^{-2}.

a If the initial velocity is 8 m s^{-1}, how long will it take to travel 6 m?

b What assumptions have you made in modelling this situation?

5 A man drives a car with a constant acceleration of 5 m s^{-2}. After 2 seconds he sees a set of traffic lights and slows down with a retardation of 2 m s^{-2}. Given that his initial velocity was 2 m s^{-1} and that he manages to stop at the traffic lights, find the distance between the traffic lights and the point where he begins to accelerate.

6 a If a jet travels in a straight line with a constant acceleration of 10 m s^{-2} and starts from rest, how long will it take to reach a speed of 500 km h^{-1}?

b How far will it travel in this time?

 7 A particle moves in a straight line with constant acceleration. At $t = 0$, $t = 4$ and $t = 8$, where t is the time measured in seconds, the particle passes points P, Q and R respectively. The distance PQ is 100 m and the velocity at P is 20 m s^{-1}.

a Find the acceleration.   **b** Find the distance QR.

8 A bicycle leaves a point O from rest and accelerates at 1 m s^{-2}. Three seconds later a car, wishing to catch the bicycle, leaves from rest from the same point O and accelerates at 8 m s^{-2}.

   a How far apart are the car and the bicycle 4 seconds after the bicycle left O?

   b After how long do they meet?

   c How far are they from O when they meet?

SKILLS CHECK **3A EXTRA** is on the CD

3.2 Motion in a vertical plane

Motion in a straight line with constant acceleration.

Any particle travelling vertically experiences a constant acceleration towards the Earth due to gravity of magnitude 9.8 m s^{-2}. Gravity can be assumed to be constant (this is a fair modelling assumption), so the equations of motion can be applied.

> **Recall:**
> Modelling assumptions in Chapter 1.

Taking downwards as positive, the particle shown has a velocity of -6 m s^{-1} and an acceleration of 9.8 m s^{-2}.

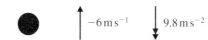

Taking upwards as positive, the particle shown has a velocity of 6 m s^{-1} and an acceleration of -9.8 m s^{-2}.

> **Note:**
> You can define either up or down as positive but be consistent within a problem.

Example 3.5 A ball is thrown upwards with speed 18 m s^{-1}.

   a Find the displacement of the particle after 1 second.

   b Find the velocity after 1 second and 2 seconds.

> **Recall:**
> Displacement is always measured from the starting point.

Step 1: Draw a clear diagram to represent the information given.

Taking upwards as positive:

> **Note:**
> Decide at the start which direction (up or down) is positive. In this case, up is positive.

Step 2: Fill the information in *suvat*, identifying what is required with a question mark.

a $s = ?$
$u = 18$
$v = $ not required
$a = -9.8$
$t = 1$
$s = ut + \frac{1}{2}at^2$
$s = (18 \times 1) + \frac{1}{2}(-9.8) \times 1^2$
$\quad = 13.1$

After 1 second the particle has a displacement of 13 m (2 s.f.).

b $s = $ not required
$u = 18$
$v = ?$
$a = -9.8$
$t = 1$ and 2
$v = u + at$
When $t = 1$
$v = 18 - (9.8 \times 1)$
$\quad = 8.2$

After 1 second the particle has a velocity of 8.2 m s^{-1}.

When $t = 2$
$v = 18 - (9.8 \times 2) = -1.6$

After 2 seconds the particle has a velocity of -1.6 m s^{-1}.

Step 3: Pick an equation of motion relating the three known variables with the unknown that is required, insert values, rearrange (if necessary) and solve.

> **Note:**
> Quote solutions to 2 s.f. when using $g = 9.8 \text{ m s}^{-1}$, unless otherwise stated.

> **Note:**
> The negative sign indicates that the particle is travelling downwards.

Example 3.6 A particle is thrown downwards in the air and moves freely under gravity. It reaches twice its initial velocity after 2 seconds.

    a Find the initial velocity of the particle.

    b Find the displacement of the particle after two seconds.

Step 1: Draw a clear diagram to represent the information given.

Taking down as positive:

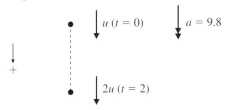

Step 2: Fill the information in *suvat*, identifying what is required with a question mark.

a
$s =$ not required
$u = ?$
$v = 2u$
$a = 9.8$
$t = 2$

b
$s = ?$
$u = 19.6$
$v = 39.2$
$a = 9.8$
$t = 2$

Step 3: Pick an equation of motion relating the three known variables with the unknown that is required, insert values, rearrange (if necessary) and solve.

$v = u + at$

$2u = u + 9.8 \times 2$

$u = 19.6$

The initial velocity of the particle is 20 m s^{-1} (2 s.f.).

$s = \dfrac{(u + v)}{2}t$

$s = \dfrac{(19.6 + 39.2)}{2} \times 2$

$= 58.8$

The displacement of the particle after 2 seconds is 59 m (2 s.f.).

The **maximum height** of a particle when it is thrown upwards, is the height at which it stops travelling up and starts falling down. At this height its velocity in the vertical direction is 0 m s^{-1}. This is an important condition when calculating maximum height reached.

Example 3.7 A cricket ball is thrown upwards at a speed of 14 m s^{-1}. By modelling the ball as a particle, find the maximum height reached and the total distance travelled when it has come back to its starting point.

Taking up as positive:

Step 1: Draw a clear diagram to represent the information given.

Step 2: Fill the information in *suvat*, identifying what is required with a question mark.

$s = ?$
$u = 14$
$v = 0$
$a = -9.8$
$t =$ not required

Step 3: Pick an equation of motion relating the three known variables with the unknown that is required, insert values, rearrange (if necessary) and solve.

$v^2 = u^2 + 2as$
$0^2 = 14^2 + 2(-9.8)s$
$s = 10$

The maximum height reached is 10 m. The total distance travelled is $2 \times 10 = 20$ m.

The **time of flight** is the total time that a particle is in the air.

Example 3.8 A girl throws a ball vertically upwards in the air with speed 15 m s^{-1} and the ball travels freely under gravity. Find:

    a the time of flight of the ball, assuming that it is thrown from the ground

    b the time for which the ball is above a height of 2 m.

Step 1: Draw a clear diagram to represent the information given.

Defining up as positive:

Step 2: Fill the information in *suvat*, identifying what is required with a question mark.

a

$s = 0$

$u = 15$

$v = $ not required

$a = -9.8$

$t = ?$

Step 3: Pick an equation of motion relating the three known variables with the unknown that is required, insert values, rearrange (if necessary) and solve.

$s = ut + \frac{1}{2}at^2$

$0 = 15t - 4.9t^2$

$0 = t(15 - 4.9t)$

$t = 0$ or $3.06...$

The time of flight is 3.1 s (2 s.f.).

b

$s = 2$

$u = 15$

$v = $ not required

$a = -9.8$

$t = ?$

$s = ut + \frac{1}{2}at^2$

$2 = 15t - 4.9t^2$

$0 = 4.9t^2 - 15t + 2$

$t = 0.140...$ or $2.92...$

The ball is above 2 m for $2.9 - 0.14 = 2.8$ s (2 s.f.).

> **Tip:**
> These are the two times for which the particle is at the given displacement.

Example 3.9 A man throws a ball vertically upwards with a speed of 15 m s^{-1} from a height of 1.5 m above the ground. The ball travels freely under gravity.

    a Calculate the time of flight.

    b Calculate the speed with which the ball hits the ground.

Step 1: Draw a clear diagram to represent the information given.

Defining up as positive:

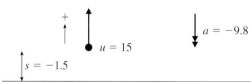

Step 2: Fill the information in *suvat*, identifying what is required with a question mark.

a

$s = -1.5$

$u = 15$

$v = $ not required

$a = -9.8$

$t = ?$

Step 3: Pick an equation of motion relating the three known variables with the unknown that is required, insert values, rearrange (if necessary) and solve.

$s = ut + \frac{1}{2}at^2$

$-1.5 = 15t - 4.9t^2$

$4.9t^2 - 15t - 1.5 = 0$

$t = 3.15...$ or $-0.096...$

The time of flight is 3.2 s (2 s.f.).

b

$s = -1.5$

$u = 15$

$v = ?$

$a = -9.8$

$t = $ not required

$v^2 = u^2 + 2as$

$v^2 = 15^2 + 2(-9.8)(-1.5)$

$v = -15.9$

The speed with which the ball hits the ground is 16 m s^{-1} (2 s.f.).

> **Note:**
> $s = 0$ at the height from which the ball is thrown and $s = -1.5$ at the ground.

> **Note:**
> The similarity between example 3.8a and example 3.9a.

> **Note:**
> The negative square root has been taken because we require the particle's velocity as it travels downwards.

Motion in a straight line with constant acceleration.

A **speed–time graph** shows how the speed of a particle varies with time. The horizontal axis represents the time taken and the vertical axis represents the speed. In the graph, u is the initial velocity and v is the final velocity after travelling for t seconds under constant acceleration.

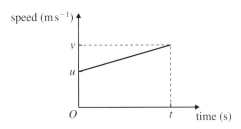

You can obtain important information from a speed–time graph by relating it to the equations of motion.

The gradient of the line $= \dfrac{(v - u)}{t} = a$.

The area under the line $= \frac{1}{2}(u + v)t = s$.

Remember, in a speed–time graph:

the gradient of the line = the acceleration of the particle
the area under the line = the distance travelled by the particle

A **velocity–time** graph shows how the velocity of a particle varies with time. In this case the area represents the displacement, and *when the area is below the time axis this represents a negative displacement.*

A **displacement–time** graph shows how the displacement of a particle varies with time. The horizontal axis represents the time taken and the vertical axis represents the displacement. We also know that:

$$\text{velocity} = \frac{\text{change in displacement}}{\text{time taken}}, \text{ and so:}$$

the gradient of a displacement–time graph
= the velocity of the particle

Recall:
From C1, the gradient, m, between two points (x_1, y_1) and (x_2, y_2) is:
$$m = \frac{y_2 - y_1}{x_2 - x_1}$$
(from $v = u + at$)

Recall:
The area of a trapezium is $\frac{1}{2}(a + b)h$, where a and b are the lengths of the parallel sides and h is the height.

Recall:
From the previous section
$$v = u + at$$
$$s = \left(\frac{u + v}{2}\right)t$$

Example 3.10 A particle travels with constant speed 4 m s^{-1} from A to B for 4 seconds. It then turns around and travels in the opposite direction from B to A with constant speed 2 m s^{-1} for a further 2 seconds.

a Sketch a velocity–time graph to represent the journey of the particle in the first 6 seconds.

b Calculate the displacement after the first 4 seconds and after 6 seconds of the motion. Hence, sketch a displacement–time graph.

Step 1: Draw a clear diagram to represent the information given.

a

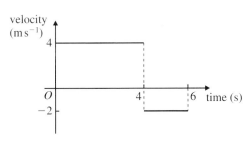

Step 2: Use area = displacement to solve the problem.

b Area under the graph for the first 4 seconds = $4 \times 4 = 16$.

So the displacement after the first 4 seconds is 16 m.

Area 'under' the graph for the last 2 seconds = $2 \times 2 = 4$.

So the displacement after 6 seconds is $16 + (-4) = 12$ m.

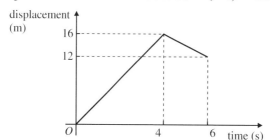

Recall:
The gradient must be constant for the displacement–time graph because the velocity is constant.

Example 3.11 A car starts from a point A and accelerates uniformly at $1\ \text{m s}^{-2}$ from rest for three seconds. It then maintains a steady speed for a further 15 seconds. Finally it slows down for 5 seconds with uniform retardation until it stops at point B.

a Sketch a speed–time graph for the journey of the car.

b Find the maximum speed of the car.

c Find the acceleration of the car in the final stage of the journey.

d Find the distance AB.

Step 1: Draw a clear diagram to represent the information given.

a Let the maximum speed be $v\ \text{m s}^{-1}$.

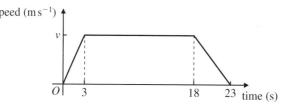

Step 2: Use gradient = acceleration and area = distance to solve the problem.

b Acceleration in first 3 seconds = gradient in first three seconds

$$1 = \frac{(v - 0)}{(3 - 0)}$$

$$v = 3$$

The maximum speed of the car is $3\ \text{m s}^{-1}$.

Note:
You could also use $v = u + at$ to solve this problem.

c Acceleration in last 5 seconds = gradient in last five seconds

$$a = \frac{0 - v}{5} = -\frac{3}{5}$$

The acceleration of the car in the last 5 seconds is $-0.6\ \text{m s}^{-2}$.

d Distance travelled from A to B = total area

$$\text{area of trapezium} = \tfrac{1}{2}(a + b)h$$
$$= \tfrac{1}{2} \times (23 + 15) \times 3$$
$$= 57$$

The distance AB is 57 m.

Note:
The area could also be calculated by dividing the trapezium into two triangles and a rectangle, and then summing their areas.

An **acceleration–time** graph shows how the acceleration of a particle varies with time. This can be sketched from a velocity–time graph by using the fact that the gradient of a velocity–time graph gives the acceleration.

Example 3.12 At $t = 0$, where t is the time in seconds, a train is travelling with a uniform velocity of 7 m s^{-1}. The train then approaches a station and, at $t = 4$, the driver applies the brake slowing down the train with uniform retardation. It reaches the station T seconds after applying the brakes, where it comes to a halt.

   a Sketch a velocity–time graph of the journey of the train.

   b Calculate the retardation, leaving your answer in terms of T.

The distance travelled by the train during the first 4 seconds is three quarters of the distance travelled during T seconds.

   c Calculate the value of T and the retardation.

   d Hence sketch an acceleration–time graph of the journey of the train.

Step 1: Draw a clear diagram to represent the information given.

a

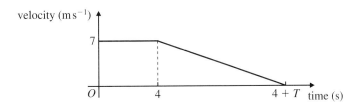

b Acceleration in last T seconds = gradient in last T seconds

Step 2: Use gradient = acceleration and area = distance to solve the problem.

$$a = \frac{(v - u)}{t} = \frac{(0 - 7)}{T} = -\frac{7}{T}$$

Hence, the retardation of the train in the last T seconds is $\dfrac{7}{T} \text{ m s}^{-2}$.

Recall:
Acceleration is given by the gradient in a speed–time graph.

c Distance travelled in first 4 s = $0.75 \times$ (distance travelled in last T s)

$$4 \times 7 = 0.75\left(\frac{1}{2} \times T \times 7\right)$$
$$T = \frac{32}{3}$$

The train slows for $10\frac{2}{3}$ s.

$$\text{Retardation} = \frac{7}{T} = \frac{21}{32} = 0.65625$$

The retardation of the train is 0.656 m s^{-2} (3 s.f.).

Recall:
Distance travelled is given by the area under a speed–time graph.

Note:
Substitute this value for T into the relation in **b**.

Recall:
The magnitude is the size.

d Acceleration–time graph:

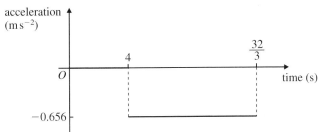

Example 3.13 A motorbike tries to catch up with a car. The car leaves 20 seconds before the motorbike, and travels with a constant speed of 8 m s^{-1}. The motorbike accelerates uniformly at 2 m s^{-2}, until it reaches a speed of 12 m s^{-1}. It then maintains this constant speed for the remainder of the journey. Find the time taken for the motorbike to reach the car.

Step 1: Draw a clear diagram to represent the information given.

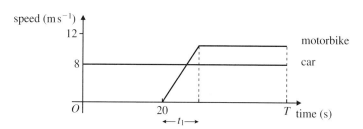

Note: The car leaves at $t = 0$, the motorbike leaves at $t = 20$ and the two meet at $t = T$.

Step 2: Use gradient = acceleration and area = distance to solve the problem.

Acceleration of motorbike = gradient of speed–time graph during t_1 s.

$$2 = \frac{12}{t_1}$$

$$t_1 = 6$$

Tip: First calculate the time for which the motorbike accelerates (t_1).

When the motorbike reaches the car

Area under line (car) = area under line (motorbike)

$$8 \times T = \tfrac{1}{2} \times 12 \times ((T - 20) + (T - 26))$$

$$T = 69$$

Note: When two particles meet, their distances, and hence the areas of their speed–time graphs, must be equal at the same given time.

The motorbike catches up with the car 69 seconds after the car left.

Hence it takes the motorbike $(69 - 20)\,\text{s} = 49\,\text{s}$ to reach the car.

SKILLS CHECK **3B: Motion in a vertical plane and graphs**

1 A ball is thrown vertically upwards with velocity $28\,\text{m s}^{-1}$ and travels freely under gravity. Find the velocity after two seconds and the distance that the particle has travelled from the start at this time.

2 Find the maximum height reached and the time of flight (the time taken to reach the start again) for the ball in question 1. Also, find for how long it is above a height of 20 m.

3 A book is dropped from 80 m above ground and travels freely under gravity. Find the time taken for it to reach the ground. What assumptions have you made when modelling this situation?

 4 A stone is thrown vertically upwards from 4 m above ground. The initial velocity of the stone is $14\,\text{m s}^{-1}$. Find:

   a the maximum height above the ground reached by the stone

   b the time taken for it to hit the ground.

5 A ball is thrown upwards and reaches a height of 50 m. Find its initial velocity.

 6 A ball is dropped from a window at $t = 0$, where t is the time in seconds. At $t = 2$ another ball is thrown from the same point with downwards velocity $25\,\text{m s}^{-1}$. Given that the balls travel freely under gravity, find:

   a the time when the balls pass each other

   b the distance from the window when they pass each other.

7 **a** Show that $h = \dfrac{u^2}{2g}$ for a particle travelling freely under gravity in the vertical plane with an initial velocity of $u\,\text{m s}^{-1}$ (upwards) and a maximum height of h m above its starting point, where g is the acceleration due to gravity.

   b Calculate the difference in the maximum heights of two particles, travelling freely under gravity, given they are thrown upwards with velocities 10 and $15\,\text{m s}^{-1}$ respectively.

8 A car leaves a point O and accelerates uniformly from rest to a speed of 4 m s^{-1} in 2 seconds. It then maintains a steady speed for a further 20 seconds, after which it slows down to a halt in 5 seconds.

    a Sketch a speed–time graph for the car's journey.

    b Find the acceleration during the initial and final stages of the journey.

    c Find the total distance travelled by the car.

9 A train travels between two stations, P and Q. The train accelerates uniformly from rest to 6 m s^{-1} in 5 seconds. It then continues its journey for a further 15 seconds with this velocity, before decelerating uniformly to a halt at station Q, in a further T seconds. The distance travelled in the last T seconds is $\frac{1}{6}$ of the total distance travelled.

    a Sketch a velocity–time graph of the motion of the train between the two stations.

    b Find T.

    c Find the distance PQ.

    d By first finding the acceleration during each stage of the journey, sketch an acceleration–time graph for the journey.

 10 Two cars set off on a journey. The first car leaves at time $t = 0$, where t is measured in seconds. It accelerates uniformly until it reaches a speed of 4 m s^{-1} at $t = 3$. It then maintains a constant velocity. The second car leaves from the same point at $t = 3$ and travels with constant speed 8 m s^{-1}.

    a On the same axes sketch speed–time graphs for the motion of the two cars.

    b Find t when the two cars meet.

    c How far are they from the start at this time?

    d Sketch a displacement–time graph for the journey of the second car.

SKILLS CHECK **3B EXTRA** is on the CD

Examination practice   Kinematics of a particle moving in a straight line

 1 A smuggler seeking to escape arrest drops a large sack containing illegal drugs from the top of a cliff which is 75 m above the surface of the sea. The sack is initially at rest and falls into the sea at the bottom of the cliff without meeting any obstacle. By modelling the sack as a particle falling with constant acceleration g, find

    a the speed with which the sack hits the sea,

    b the time taken for the sack to reach the sea.

    c Suggest two physical factors which have been ignored in modelling the situation in the above way.   [Edexcel Jan 1996]

2 A train accelerates from rest at a station A with constant acceleration 2 m s^{-2} until it reaches a speed of 36 m s^{-1}. It then cruises at this speed for 90 s before braking to stop at the next station B. The line from A to B is straight. During the braking period the train decelerates with constant deceleration 3 m s^{-2}.

    a Sketch a speed–time graph to illustrate this information.

    b Find the total time taken for the train to travel from A to B.

    c Find the distance from A to B.   [Edexcel June 1995]

3 A racing car emerging from a bend reaches a straight stretch of road. The start of the straight stretch is the point O and there are two marker points, A and B, further down the road. The distance $OA = 64$ m and the distance $OB = 250$ m. The car passes O at time 0 s and, moving with constant acceleration, passes A and B at times 2 s and 5 s respectively. Find

   a the acceleration of the car,

   b the speed of the car at B.   [Edexcel June 1996]

4 A sprinter runs a 100 metre race. He starts at a speed of 6 m s^{-1}, accelerates uniformly for 2 s to his top speed, and then maintains this top speed for the rest of the race. He covers the whole distance of 100 m in a total time of 11 s.

   a Sketch a speed–time graph to illustrate the motion of the sprinter during the time of the race.

   b Find his top speed.   [Edexcel June 1996]

5 A bus and a cyclist are moving along a straight horizontal road in the same direction. The bus starts from rest at a bus stop O and moves with a constant acceleration of 2 m s^{-2} until it reaches a maximum speed of 12 m s^{-1}. It then maintains this constant speed. The cyclist travels with a constant speed of 8 m s^{-1}. The cyclist passes O just as the bus is starting to move. The bus later overtakes the cyclist at the point A.

   a Show that the bus does not overtake the cyclist before it reaches its maximum speed.

   b Sketch, on the same diagram, speed–time graphs to represent the motion of the bus and the cyclist as they move from O to A.

   c Find the time taken for the bus and the cyclist to move from O to A.

   d Find the distance OA.   [Edexcel June 2001]

4 Statics of a particle

4.1 Resolving forces

Forces treated as vectors. Resolution of forces.

In Chapter 2, you were introduced to **resolving** vectors into their horizontal and vertical **components**. This practice can be applied to a **force**, as force is a vector quantity (it has both magnitude and direction). The exact definition of a force, measured in newtons (N), is given in the next chapter. In this chapter, you will solve problems by resolving forces into their components.

Resultant of a horizontal and vertical force

If a force $\mathbf{F}$ N has horizontal component F_h N and vertical component F_v N, as shown in the diagram, then the magnitude of the resultant can be calculated using:

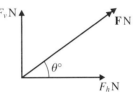

$$|\mathbf{F}| = \sqrt{(F_h^2 + F_v^2)}.$$

The angle made with the horizontal component can be calculated using

$$\tan \theta = \frac{F_v}{F_h}.$$

Recall:
You can also write this in vector form as:
$$\mathbf{F} = F_h\mathbf{i} + F_v\mathbf{j}.$$

Recall:
Another way of writing the magnitude of $\mathbf{F}$ is $F = |\mathbf{F}|$. This is also referred to as the modulus or size, F.

Example 4.1 A force of 8 N acts horizontally to the right and a force of 3 N acts vertically upwards. Find the magnitude of the resultant of these forces and the angle that the resultant makes with the horizontal force.

Step 1: Draw a force diagram resolving each force into its components.

Step 2: Find the resultant force.

Step 3: Solve for unknowns.

$$|F| = \sqrt{(F_h^2 + F_v^2)}$$
$$= \sqrt{8^2 + 3^2}$$
$$= 8.54\ldots$$

$$\tan \theta = \frac{F_v}{F_h} = \frac{3}{8}$$
$$\theta = \tan^{-1}\left(\frac{3}{8}\right) = 20.6$$

The magnitude of the resultant force is 8.54 N (3 s.f.) and the angle is 20.6°.

Note:
You may be asked to find the angle that the resultant makes with the upward vertical. This is $90° - 20.6° = 69.4°$.

Example 4.2 Find the magnitude of the resultant and the angle that the resultant makes with the positive x-axis for the force $\mathbf{F} = (3\mathbf{i} - 4\mathbf{j})$ N, where $\mathbf{i}$ and $\mathbf{j}$ are perpendicular unit vectors.

Recall:
$\mathbf{i}-\mathbf{j}$ notation.

Step 1: Draw a force diagram resolving each force into its components.

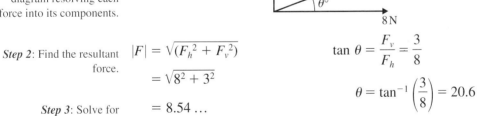

$$|\mathbf{F}| = |3\mathbf{i} - 4\mathbf{j}|$$

Step 2: Find the resultant force.

$$|F| = \sqrt{(F_h^2 + F_v^2)}$$
$$= \sqrt{(3^2 + 4^2)}$$
$$= 5$$

$$\tan \theta° = \frac{|F_v|}{F_h} = \frac{4}{3}$$
$$\theta = \tan^{-1}\left(\frac{4}{3}\right) = 53.1\ldots$$

Step 3: Solve for unknowns.

The magnitude of the resultant force is 5 N.

The angle required is below the positive x-axis, so can be written as $-53.1°$ or $306.9°$.

Tip:
Use positive values to find an angle first.

Recall:
Positive angles are measured anticlockwise from the positive x-axis.

Resolving a force into components

If a force **F** N has magnitude F N and it acts at an angle $\theta°$ to the positive x-axis you can work out its horizontal and vertical components using simple trigonometry:

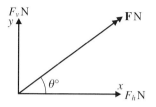

$F_h = F \cos \theta°$

$F_v = F \sin \theta°$

If **i** and **j** are perpendicular unit vectors then **F** can then be written in vector form as:

$$\mathbf{F} = F \cos \theta°\mathbf{i} + F \sin \theta°\mathbf{j}$$

Note:
Define $\theta°$ as the angle between the force and the positive x-axis.

Example 4.3 (In this question **i** and **j** are perpendicular unit vectors.) Find the horizontal and vertical components for a force of 10 N, which acts at an angle of 140° to the vector **i**. Write the force in the form $a\mathbf{i} + b\mathbf{j}$.

Step 1: Draw a force diagram resolving each force into its components.

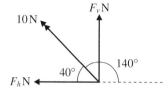

Tip:
Use the acute angle to work out the components, then incorporate a sign to give the direction.

Step 2: Find the components in each direction.

Horizontal component
$= 10 \cos 40°$ to the left
so $F_h = -7.66$ (3 s.f.)

Vertical component
$= 10 \sin 40°$ upwards
so $F_v = 6.43$ (3 s.f.)

The required force is $(-7.66\mathbf{i} + 6.43\mathbf{j})$ N (3 s.f.).

Note:
The resultant
$\sqrt{((-7.66...)^2 + 6.43...^2)} = 10$

Resultant of two or more forces

The resultant of two or more forces can be calculated by summing the vertical and horizontal components, separately.

Example 4.4 Find the resultant and the angle that the resultant makes with the 5 N force for the following system of forces.

Step 1: Draw the force diagram, and, where appropriate, redraw each force resolved into any two perpendicular directions.

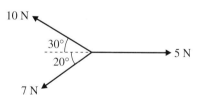

Recall:
Horizontal and vertical components are perpendicular to each other. A horizontal force has 0 component vertically.

Note:
Vectors can be resolved into any two perpendicular directions.

Step 2: Find the resultant in each direction.

Horizontal component:
$F_h = 5 - 10 \cos 30° - 7 \cos 20°$
$= -10.23...$

Vertical component:
$F_v = 10 \sin 30° - 7 \sin 20°$
$= 2.60...$

Note:
Remember to use opposite signs for forces in opposite directions.

Step 3: Find the overall resultant and the angle.

$\mathbf{F} = -10.2\mathbf{i} + 2.61\mathbf{j}$ (3 s.f.)

$F = \sqrt{(F_h^2 + F_v^2)} = 10.5$ (3 s.f.)

$\theta = \tan^{-1}\left(\frac{F_v}{|F_h|}\right) = 14.3$ (3 s.f.)

The resultant is a force of 10.5 N at an angle of 166° with the 5 N force.

Example 4.5 The following forces, **P** N, **Q** N and **R** N, act on a particle, where **i** and **j** are unit vectors acting due east and due north respectively.

$$\mathbf{P} = (3\mathbf{i} + 4\mathbf{j}), \mathbf{Q} = (2\mathbf{i} - 5\mathbf{j}), \mathbf{R} = (-\mathbf{i} - 2\mathbf{j})$$

Find the resultant force, **F**, in vector form. Also find the magnitude of the resultant and the angle that the resultant makes with the vector **i**.

Note:
There is no need to draw a vector diagram to resolve the components, as the forces are defined in **i**–**j** notation, where **i** and **j** are the horizontal and vertical unit vectors.

Step 1: Find the resultant force by adding the **i** parts and the **j** parts separately.

$$\mathbf{F} = \mathbf{P} + \mathbf{Q} + \mathbf{R} = (3\mathbf{i} + 4\mathbf{j}) + (2\mathbf{i} - 5\mathbf{j}) + (-\mathbf{i} - 2\mathbf{j})$$
$$= (3 + 2 - 1)\mathbf{i} + (4 - 5 - 2)\mathbf{j}$$
$$= (4\mathbf{i} - 3\mathbf{j})$$

Step 2: Calculate the magnitude and angle of the resultant force.

$$|\mathbf{F}| = \sqrt{(F_h^2 + F_v^2)} = \sqrt{((4)^2 + (-3)^2)} = 5$$

$$\theta = \tan^{-1}\left(\frac{|F_v|}{F_h}\right) = \tan^{-1}\left(\frac{3}{4}\right) = 36.9 \text{ (3 s.f.)}$$

The angle that the force makes with the vector **i** is $-36.9°$ (3 s.f.).

Note:
You can also write this angle as $(360° - 36.9...°) = 323°$

4.2 Equilibrium of forces

Equilibrium of a particle under coplanar forces.

A system of forces acting on a particle is in equilibrium when the resultant force is zero (there is no net force).

- **The algebraic sum of the horizontal components is 0 N.**
- **The algebraic sum of the vertical components is 0 N.**

Example 4.6 Find P and Q if the following system of forces is in equilibrium:

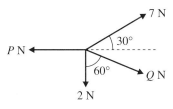

Recall:
$F_h = F\cos\theta$ and $F_v = F\sin\theta$ when θ is defined as the angle between the vector and the positive x-axis. The angle that Q makes with the vertical is 60° so the angle Q makes with the horizontal is 30°.

Step 1: Draw the force diagram, and, where appropriate, redraw each of the forces resolved into any two perpendicular directions.

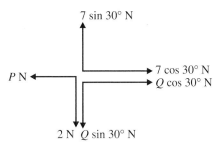

Note:
Define up and right as positive, down and left as negative.

Step 2: Find the resultant force in each direction (and equate each to zero when in equilibrium).

Step 3: Solve for unknowns.

Vertical components:

$$7\sin 30° - 2 - Q\sin 30° = 0$$

$$Q = \frac{7\sin 30° - 2}{\sin 30°} = 3$$

Force Q is 3 N.

Horizontal components:

$$7\cos 30° + Q\cos 30° - P = 0$$

$$P = 7\cos 30° + 3\cos 30°$$
$$= 8.660...$$

Force P is 8.66 N (3 s.f.).

Example 4.7 The following diagram shows a system of forces acting on a particle in a plane. A third force is added so that the particle rests in equilibrium. Find the magnitude of this force and the angle that it makes with the horizontal.

Step 1: Draw the force diagram, and, where appropriate, redraw each of the forces resolved into any two perpendicular directions.

Let the force which is added to keep the system in equilibrium be F N and let the angle it makes below the horizontal be $\theta°$.

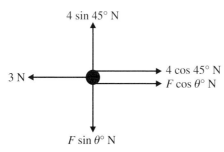

Note:
Include the unknown force and take a reasonable guess as to which direction it acts.

Step 2: Find the resultant force in each direction (and equate each to zero when in equilibrium).

Horizontal components:

$4 \cos 45° + F \cos \theta° - 3 = 0$

$F \cos \theta° = 3 - 4 \cos 45°$ (1)

Vertical components:

$4 \sin 45° - F \sin \theta° = 0$

$F \sin \theta° = 4 \sin 45°$ (2)

Step 3: Solve for unknowns.

Divide (2) by (1)

$$\tan \theta° = \frac{\sin \theta°}{\cos \theta°} = \frac{4 \sin 45°}{3 - 4 \cos 45°} = 16.48\ldots$$

$$\theta = \tan^{-1} 16.48\ldots$$

$$= 86.5 \text{ (3 s.f.)}$$

$$F = \frac{4 \sin 45°}{\sin 86.5°}$$

$$= 2.83 \text{ (3 s.f.)}$$

The added force is 2.83 N, acting at an angle of 86.5° below the horizontal.

Recall:
From C2 that $\dfrac{\sin \theta}{\cos \theta} = \tan \theta$.

When forces, defined in **i**–**j** notation, are in equilibrium then the resultant force is written as $(0\mathbf{i} + 0\mathbf{j})$ N.

Example 4.8 The forces $\mathbf{P} = (a\mathbf{i} + 4\mathbf{j})$, $\mathbf{Q} = (2\mathbf{i} - 5\mathbf{j})$ and $\mathbf{R} = (\mathbf{i} + b\mathbf{j})$, in newtons, act on a particle, where **i** and **j** are perpendicular unit vectors. The particle is at rest in equilibrium. Find a and b.

Step 1: Find the resultant force by adding the **i** parts and the **j** parts separately (and equate to $0\mathbf{i} + 0\mathbf{j}$ when in equilibrium).

$\mathbf{P} + \mathbf{Q} + \mathbf{R} = (a\mathbf{i} + 4\mathbf{j}) + (2\mathbf{i} - 5\mathbf{j}) + (\mathbf{i} + b\mathbf{j}) = 0\mathbf{i} + 0\mathbf{j}$

$(a + 2 + 1)\mathbf{i} + (4 - 5 + b)\mathbf{j} = 0\mathbf{i} + 0\mathbf{j}$

$(a + 3)\mathbf{i} + (b - 1)\mathbf{j} = 0\mathbf{i} + 0\mathbf{j}$

Tip:
Sum the **i** and **j** terms separately.

Step 2: Solve for the unknowns by equating the **i** and **j** components separately.

Equating **i** components:

$a + 3 = 0$

$a = -3$

Equating **j** components:

$b - 1 = 0$

$b = 1$

In this exercise, where applicable, **i** and **j** are perpendicular unit vectors.

1 Find the magnitude of the resultant and the angle that the resultant makes with the positive x-axis for the following forces, giving your answers to one decimal place:

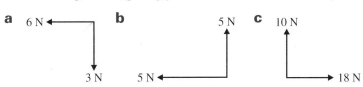

d $(3\mathbf{i} + 2\mathbf{j})\,\text{N}$ **e** $(3\mathbf{i} - 5\mathbf{j})\,\text{N}$.

2 Find the horizontal and vertical components of the following forces:

a a force of 10 N acting at 10° to the horizontal

b a force of 25 N acting at 15° to the horizontal

c a force of 35 N acting at 85° to the horizontal.

3 Find the magnitude of the horizontal and vertical components of the following forces that act on a particle, leaving your answer in the form $a\mathbf{i} + b\mathbf{j}$, where a and b are evaluated to one decimal place:

a a force of 5 N acting at 120° to the horizontal

b a force of 20 N acting at 30° below the horizontal

c a force of 145 N acting at 220° to the horizontal.

 4 The diagram shows forces **P** N, **Q** N and **R** N acting on a particle. The line of action of the force **Q** is in the horizontal direction.

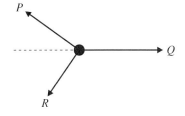

Find the magnitude of the resultant of these forces and the angle that the resultant makes with the direction of force **Q** if:

a **P** = 10 acting at 30° above the horizontal, **Q** = 7 and **R** = 15 acting at 15° below the horizontal

b **P** = 50 acting at 19° above the horizontal, **Q** = 45 and **R** = 100 acting at 18° below the horizontal.

 5 Find the magnitude of the resultant of the forces **P** N, **Q** N and **R** N and the angle that the resultant makes with the vector **i** if:

a **P** = 7**i** + 2**j**, **Q** = 3**i** + 2**j**, **R** = 5**i** – 2**j**   **b** **P** = 3**i** – 2**j**, **Q** = –3**i** –5**j**, **R** = –2**i** + 2**j**.

6 The diagram shows the forces **A** N, **B** N and **C** N acting on a particle. The forces are in equilibrium.
Find **A** and **B** when:

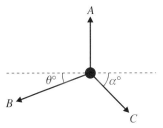

a $\theta = 25$, $\alpha = 40$, **C** = 10

b $\theta = 18$, $\alpha = 20$, **C** = 19.

Quote your answers to a suitable degree of accuracy.

 7 A particle is acted on by a force of 15 N which acts on a bearing of 020°, and another force of 4 N which acts on a bearing of 230°. Find the magnitude of a third force which will keep the system in equilibrium, stating the angle of its line of action as a bearing.

8 A particle is in equilibrium at O under the action of forces $\mathbf{P}$ N, $\mathbf{Q}$ N and $\mathbf{R}$ N. Find the values of a and b if:

a $\mathbf{P} = (a\mathbf{i} + 6\mathbf{j})$, $\mathbf{Q} = (5\mathbf{i} + b\mathbf{j})$, $\mathbf{R} = (9\mathbf{i} - 2\mathbf{j})$

b $\mathbf{P} = (2\mathbf{i} - 5\mathbf{j})$, $\mathbf{Q} = (a\mathbf{i} - 5\mathbf{j})$, $\mathbf{R} = (-3\mathbf{i} + b\mathbf{j})$

c $\mathbf{P} = (-4\mathbf{i} - 3\mathbf{j})$, $\mathbf{Q} = (2\mathbf{i} - 11\mathbf{j})$, $\mathbf{R} = (a\mathbf{i} + b\mathbf{j})$.

SKILLS CHECK **4A EXTRA** is on the CD

4.3 Types of force

Weight, normal reaction, tension and thrust, friction.

Weight is the gravitational attraction between a particle of mass m kg and the Earth. It is often written as:

$$\text{weight} = mg$$

where g is the acceleration due to gravity and has approximate value 9.8 ms^{-2}. Weight is a force that always acts vertically downwards, towards the centre of the Earth.

Normal reaction force is the force exerted by a surface on a particle which lies in equilibrium. This force always acts in a direction that is perpendicular to the surface, opposing the direction of the weight.

Tension is the resisting force provided by a string, when the string holds the particle in equilibrium. It acts away from the particle.

Thrust is the resisting force provided by a spring, when the spring holds the particle in equilibrium. It acts in a direction to oppose the compression or extension.

Friction is the force that opposes motion on a **rough** surface. It is caused by the contact between the particle and the surface. On a **smooth** surface the friction is 0 N.

These forces are used in unit M1. You can apply the techniques learnt above to questions involving these forces acting on a particle.

Example 4.9 A block of mass 3 kg rests on a rough, horizontal surface. The block is pushed with a horizontal force of 3 N. The block is kept in equilibrium by a frictional force F N. Find F and the normal reaction.

Step 1: Draw the force diagram and, where appropriate, redraw each of the forces resolved into any two perpendicular directions.

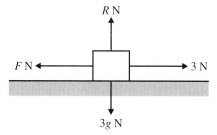

Step 2: Find the resultant force in each direction (and equate each to 0 when in equilibrium).

Horizontal components:

$F - 3 = 0$

$F = 3$

Vertical components:

$R - 3g = 0$

$R = 3g$

$= 29.4$

Step 3: Solve for unknowns.

The frictional force is 3 N.

The normal reaction is 29.4 N.

Note:
You can draw a pushing force on one side as a pulling force on the opposite side.

Recall:
The frictional force will act to oppose the pushing force of 3 N.

Note:
Equate the horizontal and vertical components to zero separately.

Recall:
$g = 9.8$

Example 4.10 A particle of mass 3 kg rests on a smooth horizontal table. A string with tension T N acts an angle 45° above the horizontal and pulls the particle. Another force of 15 N acts at an angle of 30° below the horizontal and opposes the tension in the string to keep the particle in equilibrium. Find the reaction force and the tension in the string.

Step 1: Draw the force diagram and, where appropriate, redraw each of the forces resolved into any two perpendicular directions.

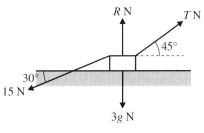

Note:
The method is the same as in the previous section. The only change is in describing the unknowns.

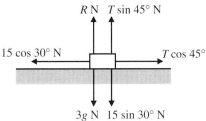

Step 2: Find the resultant force in each direction (and equate each to 0 when in equilibrium).

Step 3: Solve for unknowns.

Horizontal components:
$T \cos 45° - 15 \cos 30° = 0$

$T = \dfrac{15 \cos 30°}{\cos 45°} = 18.37\ldots$

The tension in the string is 18.4 N (3 s.f.).

Vertical components:
$R + T \sin 45° - 15 \sin 30° - 3g = 0$

$R = 15 \sin 30° + 3g - T \sin 45°$
$\quad = 23.90\ldots$

The normal reaction is 23.9 N (3 s.f.).

Sometimes a system of forces may act to suspend a particle freely in the air. In this case there will be no reaction force because the particle is not in contact with any surface.

Example 4.11 A particle of mass 2 kg is suspended freely by two strings and hangs in equilibrium as shown. Find the tension in the two strings.

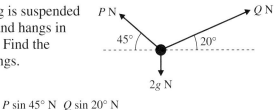

Step 1: Draw the force diagram and, where appropriate, redraw each of the forces resolved into any two perpendicular directions.

Note:
Remember to include the weight.

Step 2: Find the resultant force in each direction (and equate each to 0 when in equilibrium).

Step 3: Solve for unknowns.

Horizontal components:
$Q \cos 20° - P \cos 45° = 0$

$Q = \dfrac{P \cos 45°}{\cos 20°}$

Substitute $P = 20.32\ldots$
$Q = 15.29\ldots$

Vertical components:
$P \sin 45° + Q \sin 20° - 2g = 0$

$2g = P \sin 45° + \dfrac{P \cos 45°}{\cos 20°} \sin 20°$

$19.6 = P(\sin 45° + \dfrac{\cos 45°}{\cos 20°} \sin 20)$
$\quad\quad P = 20.32\ldots$

The tensions in strings are 20.3 N (3 s.f.) and 15.3 N (3 s.f.).

Tip:
You need to practise solving these types of simultaneous equations.

4.4 Friction and the coefficient of friction

Coefficient of friction.

When a particle of mass m kg rests on a rough surface, there is a frictional force **F** N which has a maximum value, F_{max} N. Suppose there is a pulling force of **P** N:

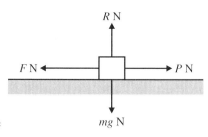

Consider the horizontal components of the forces; there are three situations that can arise:

1 If the force $P < F_{max}$ then the frictional force $F = P$, i.e. friction will take a value sufficient to maintain equilibrium.

2 If the force $P = F_{max}$ then the situation is said to be in **limiting equilibrium** and the system is at the point of moving.

3 If the $P > F_{max}$ then equilibrium is broken, the object will move and friction will maintain its maximum value, F_{max}.

When the situation is in limiting equilibrium then

$$F_{max} = \mu R$$

where μ is called the **coefficient of friction** and R is the normal reaction force of the particle. μ is a property of surface; rough surfaces have large values of μ and exert a large force of friction. Similarly, large values of the normal reaction cause large forces of friction.

> **Recall:**
> On a smooth surface, **F** = 0 N.

> **Note:**
> If there is no pulling force then the frictional force exerted = 0 N.

> **Note:**
> If the frictional force acting is F then:
> $$0 < F \leqslant F_{max}.$$

> **Recall:**
> The normal reaction force is related to the weight.

Example 4.12 A car of mass 1000 kg, is being pulled along a rough, horizontal surface by a rope. The tension in the rope is 5000 N and the rope is inclined at an angle of 15° above the horizontal. Find the normal reaction force and the coefficient of friction, given that the car is about to move.

Step 1: Draw the force diagram and, where appropriate, redraw each of the forces resolved into any two perpendicular directions.

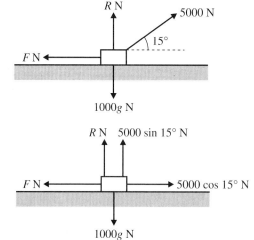

> **Note:**
> The method remains the same as in Sections 4.2 and 4.3; the only change is in the unknowns that you are asked to find.

> **Note:**
> **F** = μ**R** only in limiting equilibrium.

Step 2: Find the resultant force in each direction (and equate each to 0 when in equilibrium).

Step 3: Solve for unknowns, using $F = \mu R$ when in limiting equilibrium.

Horizontal components:

$5000 \cos 15° - F = 0$

$F = 5000 \cos 15°$

Limiting equilibrium: $F = \mu R$

$$\mu = \frac{5000 \cos 15°}{8505.9\ldots}$$

The coefficient of friction is 0.57 (2 s.f.).

Vertical components:

$R + 5000 \sin 15° - 1000g = 0$

$R = 1000g - 5000 \sin 15°$

$\quad = 8505.9\ldots$

The normal reaction is 8500 N (2 s.f.).

> **Note:**
> μ has no units.

> **Note:**
> Quote answers to 2 s.f. when using $g = 9.8$.

Forces on an inclined plane

If a particle is on an inclined plane then, instead of resolving horizontally and vertically, resolve parallel to the plane and perpendicular to the plane, as this simplifies the algebra.

Parallel and perpendicular components of the weight

If a particle of mass m kg rests on a plane inclined at angle θ to the horizontal, the weight acts vertically downwards:

Draw the force diagram:

The components parallel and perpendicular to the plane are:

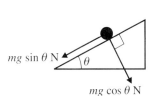

Note:

This can be seen by drawing a force triangle:

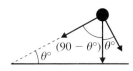

The parallel component of the weight is $mg \sin \theta$ down the plane.
The perpendicular component of the weight is $mg \cos \theta$.

Parallel and perpendicular components of the normal reaction force and friction

The normal reaction always acts perpendicular to the surface in contact so it only has a perpendicular component ($= R$). Similarly, the friction is always parallel to the plane and so it only has a parallel component, in a direction opposing relative motion.

Note:

There is no need to resolve the normal reaction force and friction.

Example 4.13 A particle of mass 4 kg rests on a smooth plane inclined at 30° to the horizontal. A force of P N acts on the particle up the plane along the line of greatest slope. Find the magnitude of the reaction force and the magnitude of the force P.

Step 1: Draw the force diagram and, where appropriate, redraw each of the forces resolved into any two perpendicular directions.

Draw the force diagram:

The components parallel and perpendicular to the plane are:

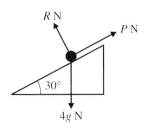

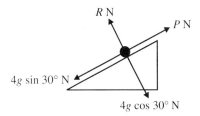

Note:

Resolve the forces in perpendicular directions (parallel and perpendicular to the plane).

Step 2: Find the resultant force in each direction (and equate each to 0 when in equilibrium).

Step 3: Solve for unknowns.

Perpendicular to the plane:

$0 = R - 4g \cos 30°$

$\therefore R = 4g \cos 30°$

$= 33.9\ldots$

Parallel to the plane:

$0 = P - 4g \sin 30°$

$P = 4g \sin 30°$

$= 19.6$

The reaction force has magnitude 34 N (2 s.f.) and the force P has magnitude 19.6 N.

1 A particle of mass m kg rests on a rough horizontal plane. A horizontal force of P N acts on the particle. Find the magnitude of the normal reaction force and the frictional force if:

a $m = 0.2, P = 7$ **b** $m = 3, P = 14$ **c** $m = 9, P = 24$.

2 A particle of mass m kg lies at rest on a rough horizontal plane and is on the point of slipping. A string applies a tension T N at an angle $30°$ above the horizontal and pulls the particle along the plane. If the frictional force is F N along the plane and the normal reaction force of the plane on the particle is R N, find the magnitude of:

a F and R if $m = 5$ and $T = 12$

b T and m if $F = 12$ and $R = 2$

c T and R if $F = 6$ and $m = 0.9$.

 3 Two men try to push a car of mass 1500 kg, which lies on a rough horizontal road. The two men apply forces of 50 N and 100 N. The car does not move. No other forces are present in the horizontal plane except friction.

a Find the frictional force between the car and the road.

b Given that the car is about to move find the coefficient of friction between the car and the road.

4 A particle of mass m kg is at rest on a rough horizontal table. A string with tension T N acting at an angle of $15°$ below the horizontal pulls the particle along the plane. The coefficient of friction between the plane and the block is μ and the normal reaction force exerted by the plane on the block is R N. Equilibrium is about to be broken; find the magnitude of:

a R and μ if $m = 2$ and $T = 20$

b T and R if $m = 0.5$ and $\mu = 0.25$

c μ and m if $T = 15$ and $R = 15$.

5 A particle of mass 5 kg is suspended freely in the air by two light strings as shown in the diagram. Find the tensions in the two strings.

 6 The diagram shows a particle suspended from a horizontal beam by two unequal, light and inextensible strings. Given that the tension in the left string is 8 N and it makes an angle of $40°$ to the beam, and the other string makes an angle of $60°$ to the beam, find the tension in the other string and the mass of the particle.

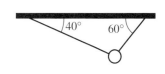

7 A particle of mass m kg lies on a rough plane, inclined at $\theta°$ to the horizontal where $\cos \theta° = \frac{12}{13}$. The system is in equilibrium. Find the magnitude of the normal reaction force and the frictional force, leaving your answer in terms of m and g, where g is the acceleration due to gravity.
Given also that the particle is in limiting equilibrium, find the coefficient of friction between the plane and the particle.

8 A particle of mass 3 kg lies on a rough plane inclined at $30°$ to the horizontal. A force of X N acts up the plane, along the line of greatest slope of the plane. The coefficient of friction between the particle and the plane is 0.5. Find the magnitude of X if:

a the particle is on the point of slipping up the plane

b the particle is on the point of slipping down the plane.

9 **a** Repeat question **8a** with the force of X N parallel to the plane replaced by a horizontal force of X N acting in the vertical plane containing the line of greatest slope of the inclined plane through the particle.

b Repeat question **8b** with the force of X N parallel to the plane replaced by a force of X N pulling the particle up the plane, acting at $45°$ above the line of greatest slope of the plane.

10 A car of mass $1500\,\text{kg}$ is broken down on a rough plane inclined at an angle of $\theta°$ to the horizontal, where $\theta° = \sin^{-1}\left(\frac{7}{25}\right)$. It is being pulled up the plane by means of a towrope, which is acting at $9°$ above the line of greatest slope of the plane. The resistance between the plane and the car have magnitude $1000\,\text{N}$. The car is at rest in equilibrium and is about to move up the plane. Find the tension in the towrope and the magnitude of the force of the plane on the car.

SKILLS CHECK **4B EXTRA** is on the CD

Examination practice   Statics of a particle

1

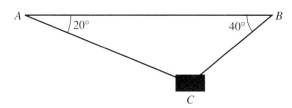

The two ends of a string are attached to two points A and B of a horizontal beam. A package of mass $2\,\text{kg}$ is attached to the string at the point C. When the package hangs in equilibrium $\angle BAC = 20°$ and $\angle ABC = 40°$, as shown above.

By modelling the package as a particle and the string as light and inextensible, find, to 3 significant figures,

a the tension in AC,   **b** the tension in BC.   [Edexcel Jan 1997]

2

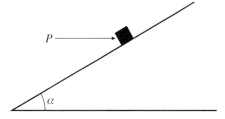

A box of mass m is placed on a plane, which is inclined at an angle α to the horizontal, where $\tan \alpha = \frac{3}{4}$. The plane is rough and the coefficient of friction between the box and the plane is $\frac{1}{2}$. The box is kept in equilibrium on the plane by applying a horizontal force of magnitude P to it, acting in a vertical plane containing a line of greatest slope of the plane, as shown above. Given that P has the smallest possible value which will enable the box to remain in equilibrium,

a draw a diagram, showing all the forces acting on the box, and indicating clearly the direction in which they act,

b find P in terms of m and g.

If instead P were to have the largest value which would enable the box to stay in equilibrium on the plane,

c state how the diagram of forces acting on the box should, if at all, be changed.

[Edexcel June 1997]

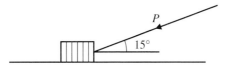

A box of mass 50 kg rests on rough, horizontal ground. The coefficient of friction between the box and the ground is 0.6. A force of magnitude P Newtons is applied to the box at an angle of 15° to the horizontal, as shown above, and the box is now in limiting equilibrium. By modelling the box as a particle find, to 3 significant figures, the value of P.   [Edexcel Jan 1999]

4

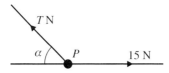

A particle P of mass 2 kg is held in equilibrium under gravity by two light, inextensible strings. One string is horizontal and the other is inclined at an angle α to the horizontal, as shown in the diagram. The tension in the horizontal string is 15 N. The tension in the other string is T newtons.

a Find the size of the angle α.

b Find the value of T.   [Edexcel Jan 2001]

 5 A particle P is acted on by three forces $\mathbf{F}_1$, $\mathbf{F}_2$ and $\mathbf{F}_3$, where $\mathbf{F}_1 = (2\mathbf{i} - 5\mathbf{j})$ N and $\mathbf{F}_2 = (4\mathbf{i} - 4\mathbf{j})$ N. Given that P is in equilibrium,

a find $\mathbf{F}_3$ in terms of $\mathbf{i}$ and $\mathbf{j}$.

The force $\mathbf{F}_3$ is now removed and P moves under the action of $\mathbf{F}_1$ and $\mathbf{F}_2$ alone.

b Find, to 3 significant figures, the magnitude of the resultant force acting on P.

c Find, in degrees to one decimal place, the angle between the resultant force acting on P and the vector $-\mathbf{j}$.   [Edexcel June 2001]

5 Dynamics of a particle moving in a straight line

5.1 Newton's laws of motion and application of F = *m*a

The concept of a force. Newton's laws of motion.

Definition of a force

A force acting on an object causes the object to accelerate. The unit of force is the **newton** (N). A force of 1 N acting on a particle of mass 1 kg causes it to accelerate at 1 m s^{-2}.

Newton's first law states that a particle will remain at rest or will continue to move with constant speed in one direction unless acted on by an external (resultant) force. In other words, a change in velocity of an object is caused by the action of a resultant force on the particle, otherwise it remains at rest or maintains constant velocity.

Newton's second law states that the resultant force, F N, produces an acceleration, **a**, that is proportional to the resultant force according to:

$$\mathbf{F} = m\mathbf{a}$$

where m is the mass of the particle.

Newton's third law states that every action has an equal and opposite reaction, i.e. if one particle applies a force on another particle, the other particle applies an equal force on the first but in the opposite direction. This is the principle behind the normal reaction force, **R** (the surface exerts an equal and opposite force on the particle that rests on the surface).

$\mathbf{F} = m\mathbf{a}$ describes motion in all possible directions. The particle will accelerate in the direction of the resultant force.

> **Note:**
> Learn this equation; remember **F** is the resultant force.

> **Recall:**
> The normal reaction R is equal to the weight of a particle, mg, on a horizontal surface when no other forces act with a vertical component.

Horizontal motion

When the force acts horizontally on an object it will accelerate in the horizontal direction according to $\mathbf{F} = m\mathbf{a}$.

Example 5.1 A particle of mass 3 kg lies on a smooth horizontal plane. A horizontal force of 18 N acts on the particle. Calculate the acceleration of the particle.

Step 1: Draw the force diagram resolving the forces into any two perpendicular directions.

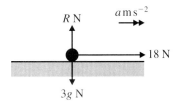

> **Note:**
> You do not need to resolve vertically in this example.

Step 2: Find the resultant force in each direction (and equate each to 0 if in equilibrium or to *m*a if accelerating).

Step 3: Solve for unknowns.

Horizontal components:

$$F_h = ma$$
$$18 = 3a$$
$$a = 6$$

Vertical components:

The particle is in equilibrium.

The particle accelerates at 6 m s^{-2} in the direction of the 18 N force.

Example 5.2 A particle of mass 5 kg is being pulled along a rough horizontal plane by a horizontal force of magnitude 15 N against a constant frictional force of magnitude 10 N. Given that the particle is initially at rest find:

a the acceleration of the particle

b the distance travelled by the particle in the first 3 seconds.

Step 1: Draw the force diagram (resolving the forces into any two perpendicular directions where necessary).

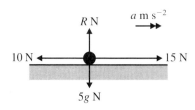

Step 2: Find the resultant force in each direction (and equate each to 0 if in equilibrium or to *ma* if accelerating).

Step 3: Solve for unknowns.

a Horizontal components:

$$F_h = ma$$

$$15 - 10 = 5a$$

$$a = 1$$

Vertical components:

The particle is in equilibrium.

The particle accelerates at 1 m s^{-2} in the direction of the 15 N force.

b $s = ut + \frac{1}{2}at^2$

$$s = 0(3) + \frac{1}{2}(1)(3)^2$$

$$s = 4.5$$

The particle travels 4.5 m in 3 s.

Example 5.3 A particle of mass 2 kg is being pulled along a rough horizontal plane by a horizontal force of magnitude 12 N. It accelerates uniformly from rest to a speed of 2 m s^{-1} in 5 seconds. Find the acceleration of the particle and hence find the coefficient of friction between the plane and the particle.

Step 1: Draw the force diagram (resolving the forces into any two perpendicular directions where necessary).

$v = u + at$

$2 = 0 + a(5)$

$a = 0.4$

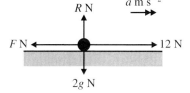

Step 2: Find the resultant force in each direction (and equate each to 0 if in equilibrium or to *ma* if accelerating).

Step 3: Solve for unknowns, using $F = \mu R$.

Horizontal components:

$F_h = ma$

$12 - F = 2(0.4)$

$F = 11.2$

$F = \mu R$

$$\mu = \frac{F}{R} = \frac{11.2}{19.6} = 0.57 \text{ (2 s.f.)}$$

Vertical components:

$F_v = 0$

$R - 2g = 0$

$R = 19.6$

The normal reaction is 19.6 N.

Vertical motion

When the net force acts vertically on an object it will accelerate in the vertical plane according to $\mathbf{F} = m\mathbf{a}$.

Example 5.4 A stone of mass 2 kg is attached to the lower end of a string hanging vertically. The particle is raised and moves with an acceleration of 5 m s^{-2}. Find the tension in the string.

Step 1: Draw the force diagram resolving the forces into any two perpendicular directions.

Note:
Let the tension in the string be **T** N.

Note:
Forces are all vertical and so do not need to be resolved into two directions.

Step 2: Find the resultant force in each direction (and equate each to 0 if in equilibrium or to *ma* if accelerating).

Step 3: Solve for unknowns.

No horizontal components.

Vertical components:

$$F_v = m a$$
$$T - 2g = 2(5)$$
$$T = 2g + 10$$
$$= 29.6$$

The tension in the string is 29.6 N.

Motion on an inclined plane

You can also apply **F** = *m***a** when a particle is on an inclined plane if there is a resultant force that will cause the particle to accelerate up or down the plane. The resultant force parallel to the plane will be referred to as $F_{||}$.

Example 5.5 A block of mass 13 kg is released from rest on a rough plane inclined at an angle of $\theta°$, where $\sin \theta° = \frac{5}{13}$. It slides down the plane and reaches a speed of 4 m s^{-1} in 2 seconds. Using the equations of motion, find the coefficient of friction between the plane and the block.

Drawing a force diagram:

The components parallel and perpendicular to the plane are:

Step 1: Draw the force diagram resolving the forces into any two perpendicular directions.

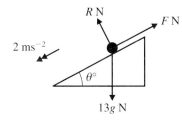

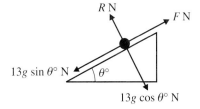

Note:
Use $v = u + at$ with $v = 4$, $u = 0$, $t = 2$.
$4 = 0 + 2a$
$a = 2$
The acceleration is 2 m s^{-2}.

Step 2: Find the resultant force in each direction (and equate each to 0 if in equilibrium or to *ma* if accelerating).

Step 3: Solve for unknowns, using $F = \mu R$.

Parallel to the plane:

$$F = ma$$
$$13g \sin \theta° - F = 13(2)$$
$$5g - F = 26$$
$$F = 5g - 26$$
$$= 23$$
$$F = \mu R$$

Perpendicular to the plane: (in equilibrium)

$$R - 13g \cos \theta° = 0$$
$$R - 12g = 0$$
$$R = 12g$$

Note:
$\sin \theta° = \frac{5}{13}$
$\cos \theta° = \frac{12}{13}$

Note:
The particle is in equilibrium perpendicular to the plane.

$$\mu = \frac{F}{R} = \frac{23}{12g} = 0.20\ldots$$

The coefficient of friction is 0.20 (2 s.f.).

Motion described in i–j notation

$\mathbf{F} = m\mathbf{a}$ can be applied to problems using $\mathbf{i}$–$\mathbf{j}$ vector notation.

Example 5.6 Forces of $(5\mathbf{i} + 4\mathbf{j})$ N, $(2\mathbf{i} - 5\mathbf{j})$ N, $(\mathbf{i} + 7\mathbf{j})$ N act on a particle of mass 5 kg, where $\mathbf{i}$ and $\mathbf{j}$ are perpendicular unit vectors.

a Find the magnitude of the acceleration.

b Find the direction in which the particle is accelerating.

Recall:
i–j notation in Chapter 2.

a $\mathbf{F} = m\mathbf{a}$

Step 1: Find the resultant force by adding the $\mathbf{i}$ parts and the $\mathbf{j}$ parts separately (and equate to $0\mathbf{i} + 0\mathbf{j}$ if in equilibrium, or to $m\mathbf{a}$ if accelerating).

$$(5\mathbf{i} + 4\mathbf{j}) + (2\mathbf{i} - 5\mathbf{j}) + (\mathbf{i} + 7\mathbf{j}) = 5\mathbf{a}$$

$$8\mathbf{i} + 6\mathbf{j} = 5\mathbf{a}$$

$$\mathbf{a} = \tfrac{1}{5}(8\mathbf{i} + 6\mathbf{j})$$

$$|\mathbf{a}| = \sqrt{\left(\frac{8}{5}\right)^2 + \left(\frac{6}{5}\right)^2} = 2$$

Step 2: Solve for unknowns.

The magnitude of the acceleration is 2 m s^{-2}.

Note:
The method is similar to previous examples, but you do not need to draw a force diagram.

b $$\tan \theta^\circ = \frac{\frac{6}{5}}{\frac{8}{5}} = \frac{3}{4}$$

$$\theta^\circ = 36.9^\circ \text{ (3 s.f.)}$$

The direction in which the particle is accelerating is at an angle of 36.9° to the vector $\mathbf{i}$.

Note:
Let θ be the angle between the vector and the x-axis.

SKILLS CHECK **5A: Newton's laws of motion and application of F = ma**

1 Find the acceleration produced when a particle of mass m kg is acted on by a resultant horizontal force of F N when:

a $F = 8, m = 2$ **b** $F = 12, m = 6$ **c** $F = 0.5, m = 0.1$ **d** $F = \frac{1}{2}, m = 0.9$.

2 Find the resultant horizontal force that acts on a particle of mass m kg to produce an acceleration of a m s^{-2} when:

a $m = 3, a = 2$ **b** $m = 0.55, a = 3$ **c** $m = 0.3, a = 0.9$.

3 A particle of mass 3 kg is being pulled across a rough horizontal plane by a horizontal force of 10 N. The coefficient of friction between the particle and the plane is 0.1. Find the magnitude of the acceleration of the particle. Given also that the particle is initially at rest, find the distance moved by the particle in the first 4 seconds.

4 A truck of mass 1500 kg is brought to rest in 5 seconds from a speed of 15 m s^{-1} on a smooth horizontal plane. Neglecting air resistance, find the braking force required to achieve this.

 5 The driving force produced by the engine of a car of mass one tonne causes a car to accelerate uniformly from rest to a speed of 8 m s^{-1} in 6 seconds along a rough horizontal road. If the coefficient of friction between the car and the plane is 0.3, find the driving force of the car.

6 A wall is raised vertically upwards, by a rope. It starts from rest and travels a vertical distance of 4 m in 6 seconds. If the tension in the rope is 400 N, find the mass of the wall. What assumptions have you made when modelling this situation?

7 A boy of mass 30 kg slides down a smooth plane inclined 45° to the horizontal. Find the acceleration of the boy.

8 Repeat question **5**, with the car moving up a rough plane inclined 30° to the horizontal.

 9 A particle of mass 6.5 kg is in limiting equilibrium on a rough plane inclined at $\theta°$ to the horizontal where $\sin \theta° = \frac{5}{13}$. Find the coefficient of friction between the plane and the particle. A horizontal force of 78 N is now applied to the particle so that it starts to accelerate up the plane. Find the acceleration of the particle and hence find how far up the plane it moves in 2 seconds.

10 Forces of $(10\mathbf{i} + 2\mathbf{j})$ N, $(\mathbf{i} - 4\mathbf{j})$ N, $(2\mathbf{i} - 7\mathbf{j})$ N act on a particle of mass 5 kg. Find the magnitude of the acceleration of the particle and also find the angle that the acceleration makes with the vector $\mathbf{i}$.

11 The forces $\mathbf{P}$, $\mathbf{Q}$ and $\mathbf{R}$ (in newtons) act on a particle of mass 0.5 kg and produce an acceleration of $(\mathbf{i} - 4\mathbf{j})$ m s^{-2}.
$\mathbf{P} = (a\mathbf{i} - 4\mathbf{j})$, $\mathbf{Q} = (2\mathbf{i} + 4\mathbf{j})$, $\mathbf{R} = (-5\mathbf{i} + b\mathbf{j})$
Find the values of the constants a and b.

SKILLS CHECK **5A EXTRA** is on the CD

5.2 Connected particles and pulleys

Simple applications including the motion of two connected particles.

Horizontal motion of connected particles

When a particle is connected to another particle by a string then you can analyse the motion of the two **connected** particles by using Newton's third law. Both particles will experience the same magnitude of force, T N, but in opposite directions.

To analyse this motion apply $\mathbf{F} = m\mathbf{a}$ to each particle, separately.

Note:
Because they are connected they will both have the same acceleration.

Recall:
What assumptions must be made?

Example 5.7 Two particles of masses 5 kg and 7 kg are connected by an inextensible string. The particle of mass 7 kg is being pulled by a horizontal force of 70 N along a rough, horizontal surface. Given that the coefficient of friction between each particle and the surface is 0.25, find the acceleration of the system and the tension in the string. What assumption have you made about the string?

Note:
Define the direction of acceleration as positive.

Step 1: Draw the force diagram resolving the forces into any two perpendicular directions.

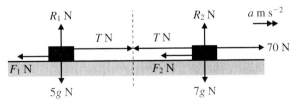

Note:
Each particle will have a different normal reaction because their masses are different.

Step 2: Find the resultant force in each direction (and equate to 0 if in equilibrium, to ma if accelerating).

5 kg particle:
$F_v = 0$ (in equilibrium)
$R_1 - 5g = 0$ ...(1)
$F_h = ma$
$T - F_1 = 5a$ ...(2)

7 kg particle:
$F_v = 0$ (in equilibrium)
$R_2 - 7g = 0$ ...(3)
$F_h = ma$
$70 - T - F_2 = 7a$ ...(4)

Note:
Treat each particle separately, as if separated by the broken line in the diagram.

Step 3: Solve for unknowns, using $F = \mu R$.

$R_1 = 49$ ...from (1) $R_2 = 68.6$ ...from (3)

$F_1 = \mu R_1 = 0.25(49) = 12.25$ $F_2 = \mu R_2 = 0.25(68.6) = 17.15$

Substitute into (2)   Substitute into (4)

$T - 12.25 = 5a$ ...(5) $70 - T - 17.15 = 7a$ ...(6)

Add equations (5) and (6): $40.6 = 12a$

$$a = 3.383... = 3.4 \text{ (2 s.f.)}$$
$$T = 5a + 12.25 \qquad \text{...from (5)}$$
$$= 29.166... = 29.2 \text{ (2 s.f.)}$$

> **Note:**
> There is no vertical acceleration (no net vertical force).

> **Note:**
> Add equations (5) and (6) to eliminate T and solve for a.

Both particles accelerate at 3.4 m s^{-2} with a tension of 29 N in the connecting string.

We have assumed that the string is light, otherwise there would have to be a vertical component for the weight of the string.

Vertical motion of connected particles

This method also applies to particles that are connected vertically.

Example 5.8   A light, inextensible string connects two bricks of equal mass 5 kg, one above the other. The system is lowered by a tow bar, which is attached to the topmost brick. It takes 10 seconds for the bricks to travel a vertical distance of 15 m, starting from rest. Find the acceleration of the bricks and use this to find the tensions in the string and the tow bar.

Step 1: Draw the force diagram resolving the forces into any two perpendicular directions.

Use $s = ut + \frac{1}{2}at^2$

$$15 = 0(10) + \tfrac{1}{2}a(10)^2$$
$$a = 0.3$$

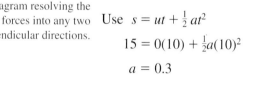
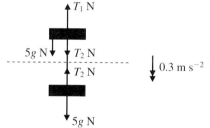

> **Note:**
> Let the tension in the tow bar be T_1, and the tension in the string be T_2.

> **Note:**
> Define the direction of acceleration (down) as positive.

Step 2: Find the resultant force in each direction (equate to 0 if in equilibrium, or to ma if accelerating).

Top brick:
$$F_v = ma$$
$$T_2 + 5g - T_1 = 5(0.3) \quad ...(1)$$
No horizontal components.

Bottom brick:
$$F_v = ma$$
$$5g - T_2 = 5(0.3) \quad ...(2)$$
No horizontal components.

Step 3: Solve for unknowns.

$$T_2 = 5g - 1.5 \qquad \text{...from (2)}$$
$$= 47.5$$
$$T_1 = T_2 + 5g - 1.5 \qquad \text{...from (1)}$$
$$= 95$$

> **Note:**
> Alternatively equations (1) and (2) could be solved simultaneously.

The tension in the tow bar is 95 N and the tension in the string is 47.5 N.

Particles connected by pulleys

The motion of two particles that are connected by a light, inextensible string passing over a smooth fixed pulley can be similarly analysed. **F** = m**a** is applied separately to each particle. As the particles are connected by a light, inextensible string the magnitude of the acceleration is the same for both particles (but it acts in opposite directions). As the pulley is smooth, the tension in the string is the same throughout the string.

Example 5.9   Particles of mass m kg and $3m$ kg are connected by a light, inextensible string, which passes over a smooth fixed pulley. Find, in terms of g, the acceleration of the system and the force exerted on the pulley.

Given that the system is released from rest, find the distance moved by one of the particles in 3 s, assuming the particle does not reach the pulley.

Step 1: Draw the force diagram resolving the forces into any two perpendicular directions.

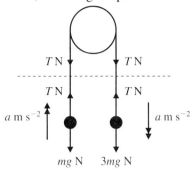

Note:
Remember to imagine a line half way down the string as shown.

Note:
Mass m moves up whilst mass $3m$ moves down.

Step 2: Find the resultant force in each direction (equate to 0 if in equilibrium, or to ma if accelerating).

Step 3: Solve for unknowns.

For mass m:

$T - mg = ma$ ...(1)

No horizontal components.

For mass $3m$:

$3mg - T = 3ma$ ...(2)

No horizontal components.

Note:
Define the direction of acceleration as positive.

(1) + (2):

$$2mg = 4ma$$

$$a = \frac{2g}{4} = \frac{g}{2}$$

The acceleration of the system is $\frac{g}{2}$ m s^{-2}.

The force exerted on the pulley is $2T$ N, where

$$T = ma + mg \qquad \text{...from (1)}$$

$$= \frac{3mg}{2}$$

So, the force exerted on the pulley is $3mg$ N downwards.

For the distance travelled in 3 seconds use $s = ut + \frac{1}{2}at^2$

$$s = 0(3) + \frac{1}{2}\left(\frac{g}{2}\right)3^2 = 22.1 \ (3 \text{ s.f.})$$

So, the distance travelled is 22.1 m.

Recall:
Equations of motion in Chapter 3.

Horizontal and vertical motion of particles connected by pulleys

Sometimes a pulley can separate two particles, one of which is resting on a plane and the other hanging freely.

Example 5.10   Two particles A and B of masses 0.5 kg and 0.7 kg respectively are connected by a light, inextensible string. Particle A lies on a rough horizontal table 8 m from a smooth peg at the edge of the table. The string passes over the peg and particle B hangs freely 2 m from the ground. The coefficient of friction between particle A and the horizontal surface is 0.2. The system is released from rest. Find:

a   the acceleration of the system

b   the time taken for B to reach the ground

c   the distance that A travels along the table, after B reaches the ground, before it comes to rest.

Note:
A smooth peg can be treated in exactly the same way as a smooth pulley.

Step 1: Draw the force diagram resolving the forces into any two perpendicular directions.

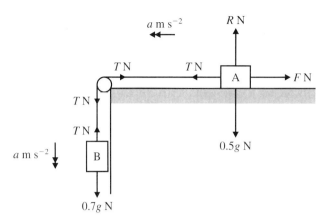

Note:
Apply $\mathbf{F} = m\mathbf{a}$ horizontally for particle A and vertically for particle B. Define the direction of motion as positive.

Step 2: Find the resultant force in each direction (equate to 0 if in equilibrium, or to ma if accelerating).

For 0.5 kg mass:
Horizontal motion:

$$F_h = ma$$
$$T - F = 0.5a \qquad \ldots(1)$$

Vertical motion:

$$F_v = 0 \text{ (in equilibrium)}$$
$$R - 0.5g = 0 \qquad \ldots(2)$$

For 0.7 kg mass:
No horizontal motion.

Vertical motion:

$$F_v = ma$$
$$0.7g - T = 0.7a \qquad \ldots(3)$$

Note:
Consider motion whilst particle B hangs freely.

Step 3: Solve for unknowns using $F = \mu R$.

a $R = 4.9$ $\qquad\qquad\qquad\qquad\qquad\qquad\qquad$ …from (2)

$F = \mu R = 0.98$

Substitute into (1) $\quad T - 0.98 = 0.5a \qquad\qquad\qquad\qquad \ldots(4)$

$$5.88 = 1.2a$$
$$a = 4.9$$

The acceleration of the particles as B falls is 4.9 m s^{-2}.

Note:
Substitute value for R into (1) then add (3) and (4).

b $\quad s = ut + \frac{1}{2}at^2 \qquad\qquad$ (where $s = 2$, $u = 0$, $a = 4.9$, $t = ?$)

$2 = 0(t) + \frac{1}{2}(4.9)t^2$

$t = \sqrt{\dfrac{4}{4.9}} = 0.903\ldots$

Particle B falls for 0.90 s (2 s.f.) before hitting the ground.

Note:
Use the value of a to calculate the time and final velocity while the string is taut.

Step 1: Find the final velocity of the particle while the string is taut.

c Find the velocity of the particles as B hits the ground:

$v^2 = u^2 + 2as \qquad\qquad$ (where $s = 2$, $u = 0$, $v = ?$, $a = 4.9$)

$v^2 = 0(2) + 2(4.9)(2) = 19.6$

$v = 4.427\ldots$

Find the acceleration of particle A:

Step 2: Find the new acceleration of the particle when there is no tension.

You can apply $\mathbf{F} = m\mathbf{a}$ to particle A or you can recognise that the equation will be the same as the one you did earlier but without any tension. So, you can use equation (4) but remove the tension to get the new acceleration when the particles are no longer connected:

$-0.98 = 0.5a \qquad\qquad\qquad\qquad\qquad\qquad\qquad$ …from (4)

$a = -1.96$

Note:
Consider A after B hits the ground.

Note:
The negative sign means that the particle is decelerating.

Step 3: Use the equations of motion to find the further distance moved by the particle until it stops.

Find the further distance travelled by A:

$v^2 = u^2 + 2as \qquad\qquad$ (where $u = 4.427\ldots$, $v = 0$, $a = -1.96$)

$0^2 = 4.427\ldots^2 + 2(-1.96)s$

$s = 5$

Particle A travels 5 m along the table after particle B hits the ground (the total distance travelled by A is $(5 + 2)$ m $= 7$ m).

Note:
The particle now decelerates from the velocity found in Step 1 to 0 m s^{-1} under this new acceleration.

Example 5.11 Two particles P and Q, of mass 10 kg and 15 kg respectively, are connected by a light, inextensible string which passes over a light, smooth pulley. Particle P rests on a smooth plane inclined at $\theta°$ to the horizontal, where $\sin\theta° = \frac{3}{5}$. Particle Q hangs vertically on the edge of the plane, 2 m above a horizontal plane. Find:

a the acceleration of the system

b the tension in the string

c the total distance that P travels up the plane, given that the string breaks after Q has travelled 1 m.

Step 1: Draw the force diagram resolving the forces into any two perpendicular directions.

Draw a force diagram:

Resolve parallel to the plane for P and vertically for Q:

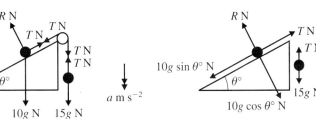

Step 2: Find the resultant force in each direction (and equate to 0 if in equilibrium, to ma if accelerating).

Step 3: Solve for unknowns.

Particle P:

Parallel to plane:
$$F_{\parallel} = ma:$$
$$T - 10g\sin\theta = 10a$$
$$T - 6g = 10a \qquad ...(1)$$

Particle Q:

Vertical motion:
$$F_v = ma$$
$$15g - T = 15a \qquad ...(2)$$

a $9g = 25a \Rightarrow a = \dfrac{9 \times 9.8}{25} = 3.528$

The acceleration of the system is 3.5 m s^{-2} (2 s.f.).

b $T = 10a + 6g = 94.08$

The tension in the string is 94 N (2 s.f.).

Step 1: Find the final velocity of the particle while the string is taut.

c Find the velocity as the string breaks:
$$v^2 = u^2 + 2as \qquad \text{(where } s = 1, u = 0, v = ?, a = 3.5)$$
$$v^2 = 0^2 + 2 \times \frac{9g}{25} \times 1$$
$$v = \sqrt{7.056}$$

Step 2: Find the new acceleration of the particle when there is no tension.

Resolve the forces on P, parallel to the plane, after the string breaks:
$$-6g = 10a$$
$$a = -5.88$$

The acceleration of particle P after the string breaks is -5.88 m s^{-2}.

Step 3: Use the equations of motion to find the further distance moved by the particle until it stops.

Find the distance moved after the string breaks:
$$v^2 = u^2 + 2as \qquad \text{(where } u = \sqrt{7.056}, v = 0, s = ?, a = -5.88)$$
$$0^2 = 7.056 - 2 \times 5.88 \times s$$
$$s = 0.6$$

So the total distance travelled by P up the plane is 1.6 m.

1 Two particles of mass 5 kg and 8 kg are on a rough horizontal surface and are connected by a light, inextensible string. The 8 kg mass is being pulled by a horizontal force of 39 N. The resistance to each particle is k times their mass, so the resistance experienced by the 5 kg mass is $5k$ N. If the acceleration of the system is 1 m s^{-2}, find the tension in the connecting string and the magnitude of the resistance experienced by each particle.

 2 A dog of mass 6 kg is attached to a sleigh of mass 3 kg by a rope and pulls it along a smooth horizontal surface. The dog exerts a forward force of F N. The dog and the sleigh start from rest and travel a distance of 15 m in 10 seconds. Find the force with which the dog pulls the sleigh and find the tension in the rope. The dog and the sleigh now reach a smooth hill inclined at $1°$ to the horizontal and continue along the line of greatest slope of the plane. If the dog maintains the same pulling force find the acceleration of the dog and the sleigh up the plane. What assumptions have you made when modelling this situation?

3 A light, inextensible string connects two particles A and B. The particles hang vertically with particle B below particle A. Particle A has a mass of 3 kg and particle B has a mass of 2 kg. Particle A is pulled upwards by means of another string with tension 85 N. Find the acceleration of the system and the tension in the lower string.

 4 A crane lowers two cars that are connected by a chain. The topmost car has mass 1000 kg and is suspended by a rope, which is connected to the crane. The lower car has mass 800 kg and is 100 m vertically above the ground. Given that it takes 10 seconds for the lower car to reach the ground, starting from rest, find the tensions in the chain and in the rope. What assumptions have you made about the chain and the rope?

5 Particles of mass 3 kg and 5 kg are attached by a light, inextensible string, which passes over a smooth fixed pulley. The 5 kg particle is 3 m above ground. The system is released from rest. By finding the acceleration of the system, find how long it takes the 5 kg mass to reach the ground. Also, find the force exerted on the pulley.

6 Particles P and Q are attached by a string that passes over a smooth fixed pulley. Particle P has twice the mass of particle Q. They both hang 2 m above horizontal ground. The system is released from rest. Find the magnitude of the acceleration of the system. Hence, find the velocity with which P hits the ground. Assuming that particle Q does not reach the pulley, find the greatest height that Q reaches above the ground. How have you used the fact that the pulley is smooth? How have you used the fact that the string is light **and** inextensible?

7 Two particles A and B are connected by a light, inextensible string. Particle A has mass 6 kg and lies on a rough horizontal table. The string passes over a smooth fixed pulley at the edge of the table and particle B of mass 5 kg hangs vertically at the other end of the string, 2 m above horizontal ground. The system is released from rest and it takes 5 seconds for particle B to hit the ground. Find the coefficient of friction between the plane and particle A.

 8 A particle P of mass 2.5 kg which is at rest on a smooth inclined plane of angle $30°$ is connected to particle Q of mass 3 kg by a light, inextensible string which lies along a line of greatest slope of the plane and passes over a fixed smooth pulley at the top of the plane. Particle Q hangs freely 1.5 m above horizontal ground. The system is released from rest with the string taut. Find:

a the acceleration of the system

b the tension in the string

c the final velocity of Q when it hits the ground

d the total distance that P moves up the plane, given that it does not reach the pulley.

9 The diagram shows two particles P and Q of masses 0.4 kg and 2 kg respectively. Particle P rests on a rough plane inclined at angle 30° to the horizontal and is attached to particle Q by means of a light, inextensible string which passes over a smooth fixed pulley at the top of the plane as shown. Particle Q lies on a smooth plane inclined at 60° to the horizontal. The coefficient of friction between particle P and the plane is $\dfrac{1}{\sqrt{3}}$. The system is released from rest.

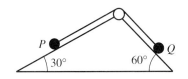

Find the acceleration of the system, given that Q travels down the plane.

SKILLS CHECK **5B EXTRA** is on the CD

5.3 Momentum and impulse

Momentum and impulse. The impulse–momentum principle. The principle of conservation of momentum applied to two particles colliding directly.

Momentum

The momentum of a particle of mass m kg, travelling with velocity $\mathbf{v}$ m s^{-1} has magnitude $m\mathbf{v}$. Momentum is a vector quantity and is measured in units of N s (newton seconds).

Momentum $= m\mathbf{v}$ N s

If the particle has speed v m s^{-1} then the momentum is given by:

Momentum $= mv$ N s

Recall:
$1\,\text{N} = 1\,\text{kg m s}^{-2}$.

Example 5.12   A particle of mass 3 kg is travelling with velocity of 4 m s^{-1}. Calculate:

a   the momentum of the particle

b   the change in momentum if the velocity of the particle increases to 8 m s^{-1}.

Step 1: Calculate unknowns using the definition of momentum.

a $mv = 3 \times 4 = 12$
The initial momentum of the particle is 12 N s.

b $mv = 3 \times 8 = 24$
The final momentum is 24 N s.
The change in momentum $= 24 - 12 = 12$ N s.

Note:
You must consider the direction of momentum.

Example 5.13   A particle of mass 3 kg has initial velocity of 2.5 m s^{-1}. Find the change in momentum of the particle if:

a   the final velocity of the particle is 6 m s^{-1}

b   the final velocity of the particle is -6 m s^{-1}.

Note:
A negative sign indicates opposite direction.

Step 1: Draw a diagram of the initial and final situations

a   Initial motion:

| 3 kg | → 2.5 m s^{-1} |

Final motion:

| 3 kg | → 6 m s^{-1} |

Step 2: Calculate unknowns using the definition of momentum.

Initial momentum $= mv$
$= 3 \times 2.5$
$= 7.5$

Final momentum $= mv$
$= 3 \times 6$
$= 18$

Change in momentum $= 18 - 7.5 = 10.5$ N s.

Step 1: Draw a diagram of the initial and final situations

b Initial motion:

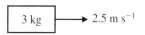

3 kg → 2.5 m s⁻¹

Final motion:

6 m s⁻¹ ← 3 kg

Note:
The direction is important. The velocity to the right is taken as positive.

Step 2: Calculate unknowns using the definition of momentum.

Initial momentum $= mv$
$$= 3 \times 2.5$$
$$= 7.5$$

Final momentum $= mv$
$$= 3 \times -6$$
$$= -18$$

Change in momentum $= (7.5 - (-18))\,\mathrm{N\,s} = 25.5\,\mathrm{N\,s}.$

Impulse

The impulse of a constant force **F** N acting for a time t seconds on a particle has magnitude Ft. Impulse has units of N s (newton seconds). An impulse changes the velocity of a particle upon which a force acts. You can calculate the change in the velocity, using the equations of motion for constant acceleration and $\mathbf{F} = m\mathbf{a}$:

$$\text{Impulse} = \mathbf{F}t$$
$$= (m\mathbf{a})t$$
$$= m\left(\frac{\mathbf{v} - \mathbf{u}}{t}\right)t$$
$$= m\mathbf{v} - m\mathbf{u}$$

In other words, the impulse of a force is the final momentum minus the initial momentum.

Recall:
$\mathbf{v} = \mathbf{u} + \mathbf{a}t$ can be written as:
$$\mathbf{a} = \frac{\mathbf{v} - \mathbf{u}}{t}$$

Note:
Impulse = change in momentum.

Example 5.14 A force of 10 N acts on a particle of mass 5 kg for 3 seconds. The particle is initially at rest. Find:

a the magnitude of the impulse given to the particle by the force

b the final velocity of the particle.

Step 1: Draw a diagram for the initial and final situations.

a Initial motion:

5 kg → 10 N
0 m s⁻¹

Final motion:

5 kg → v m s⁻¹

Note:
Let final velocity $= v$ m s⁻¹.

Step 2: Calculate unknowns using the definition of momentum and impulse.

Initial momentum $= mv$
$$= 5 \times 0$$
$$= 0$$

Final momentum $= mv$
$$= 5 \times v$$
$$= 5v$$

a Impulse $= Ft = 10 \times 3 = 30$
The impulse on the particle is 30 N s.

b Impulse $= mv - mu$
$$30 = 5v - 0$$
$$v = 6$$
The final velocity of the particle is 6 m s⁻¹.

Recall:
$Ft = m\mathbf{v} - m\mathbf{u}$

Example 5.15 A ball of mass 15 kg collides at right angles with a fixed vertical wall at a velocity of 4 m s⁻¹. The ball rebounds with velocity of -2 m s⁻¹. Find the magnitude of the impulse exerted by the wall on the ball.

Step 1: Draw a diagram for the initial and final situations.

a Initial motion:

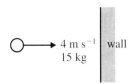

4 m s^{-1} wall
15 kg

Final motion:

2 m s^{-1} ⟵ ◯ wall
15 kg

Step 2: Calculate unknowns using the definition of momentum and impulse.

Initial momentum $= mv$

$$= (15)(-4)$$
$$= -60$$

Final momentum $= mv$

$$= (15)(2)$$
$$= 30$$

Note:
Let motion to the left be positive.

Impulse $= mv - mu$

$$= 30 - (-60)$$
$$= 90$$

The impulse exerted by the wall on the ball has magnitude 90 Ns.

Momentum and impulse in i–j notation

Both momentum and impulse are vector quantities because they depend on the velocity of a particle. Sometimes questions will involve the use of vector notation when calculating momentum and impulse:

$$\text{Momentum} = m\mathbf{v}$$

$$\text{Impulse:} \quad \mathbf{I} = \mathbf{F}t = m\mathbf{v} - m\mathbf{u}$$

Example 5.16 A particle of mass 500 g has an initial velocity of $(2\mathbf{i} - 4\mathbf{j})$ m s^{-1}, where $\mathbf{i}$ and $\mathbf{j}$ are unit vectors perpendicular to each other.
A force $\mathbf{F}$ N acts on the particle for 3 seconds and changes the velocity of the particle to $(-16\mathbf{i} + 44\mathbf{j})$ m s^{-1}. Find $\mathbf{F}$ in the form $(a\mathbf{i} + b\mathbf{j})$.

Recall:
The mass has to be in kilograms when calculating momentum.

Step 1: Calculate unknowns using the definition of momentum and impulse.

$$\text{Impulse} = \mathbf{F}t$$

where

$$\mathbf{F}t = m\mathbf{v} - m\mathbf{u}$$

$$3(a\mathbf{i} + b\mathbf{j}) = 0.5(-16\mathbf{i} + 44\mathbf{j}) - 0.5(2\mathbf{i} - 4\mathbf{j})$$

$$= -9\mathbf{i} + 24\mathbf{j}$$

$$a\mathbf{i} + b\mathbf{j} = -3\mathbf{i} + 8\mathbf{j}$$

The force acting on the particle is $(-3\mathbf{i} + 8\mathbf{j})$ N.

The principle of conservation of momentum

Momentum is conserved between two colliding particles when there are no external forces acting. That is, **the total momentum before the collision equals the total momentum after the collision.** You can use this relation to solve problems involving colliding particles.

Example 5.17 A particle of mass 4 kg travels horizontally with a velocity of 8 m s^{-1}. It collides with another particle of mass 3 kg, which is at rest. After the impact the 4 kg mass has a velocity of -5 m s^{-1}. Find the velocity of the 3 kg mass after the impact.

Note:
Let the speed of the 3 kg mass after the collision be v m s^{-1}.

Step 1: Draw a diagram for the initial and final situations.

Before collision:

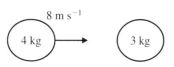

After collision:

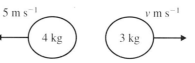

Note:
The direction of motion is very important, so choose which direction is positive at the start and stick with it.

Step 2: Calculate the total momentum before and after the collision.

Momentum before collision
$$= 4 \times 8 + 0$$
$$= 32$$

Momentum after collision
$$= 4 \times (-5) + 3v$$
$$= -20 + 3v$$

Note:
Let motion from left to right be positive.

Step 3: Calculate unknowns using the principle of conservation of momentum.

By the principle of conservation of momentum
$$32 = -20 + 3v$$
$$v = \tfrac{52}{3} = 17.33\ldots$$

The final velocity of the particle is $17\ \text{m s}^{-1}$ (2 s.f.).

Recall:
Momentum before
= momentum after.

Example 5.18 Two particles A and B of masses 2 kg and 1 kg respectively are moving towards each other in the same straight line with speeds $2u\ \text{m s}^{-1}$ and $u\ \text{m s}^{-1}$, respectively. After the impact, the particles coalesce and continue to travel in the direction of particle A before impact, with speed $5\ \text{m s}^{-1}$. Find the initial speeds of A and B.

Note:
When two particles coalesce, they join together and you can treat them as a single particle.

Step 1: Draw a diagram for the initial and final situations.

Before collision:

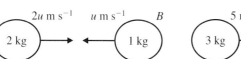

After collision:

Note:
Motion from left to right is positive.

Step 2: Calculate the total momentum before and after the collision.

Momentum before collision
$$= 2(2u) - u$$
$$= 3u$$

Momentum after collision
$$= 15$$

Tip:
It is useful to simplify each equation at this stage of the calculation.

Step 3: Calculate unknowns using the principle of conservation of momentum.

By the principle of conservation of momentum
$$3u = 15$$
$$u = 5$$

The initial speeds of A and B are $10\ \text{m s}^{-1}$ and $5\ \text{m s}^{-1}$, respectively.

Gun firing a bullet

When a gun fires a bullet the initial momentum is 0 N s, assuming there is no initial movement. But on firing the bullet there is a **recoil** in the gun which is in the opposite direction to the motion of the bullet, according to the conservation of momentum.

Recall:
Initial momentum = final momentum.

Example 5.19 A bullet is fired by a gun, which is 3 kg heavier than the bullet. Given that, after the shot is fired, the bullet travels in a straight line with velocity $200\ \text{m s}^{-1}$ and the gun recoils in the opposite direction with velocity $5\ \text{m s}^{-1}$, find the mass of the bullet and the mass of the gun.

Step 1: Draw a diagram for the initial and final situations.

a

b

$5\ \text{m s}^{-1}$

$200\ \text{m s}^{-1}$

Note:
Let the mass of the bullet be m kg. Then the mass of the gun is $(m + 3)$ kg.

Step 2: Calculate the total momentum before and after the collision.

Momentum before firing	Momentum after firing
$= 0\,\text{N s}$	$= (3 + m)(-5) + 200m$
	$= (195m - 15)\,\text{N s}$

Note:
Motion from left to right is positive.

Step 3: Calculate unknowns using the principle of conservation of momentum.

By the principle of conservation of momentum
$$0 = 195m - 15$$
$$m = 0.0769\ldots$$

So, the mass of the bullet is 77 g (2 s.f.) and the mass of the gun is 3.08 kg (3 s.f.).

Note:
Because the value of m is so small it is better to convert it to grams.

$$1\,\text{kg} = 1000\,\text{g}$$

Jerk in a string

If two particles are connected by a (light) string, which is initially slack, and one particle is given an initial velocity, then when the string becomes taut there will be a jerk in the string. After the jerk, **both particles will move off with the same velocity.**

Note:
Let the particle's velocity be $v\,\text{m s}^{-1}$.

Example 5.20 Two identical particles P and Q, each of mass m kg, lie on a smooth, horizontal table. They are connected at either end of a light, inextensible string which initially is slack. Particle Q is projected away from P with an initial velocity of 8 m s^{-1}. Find:

a the common velocity of the particles after the string jerks

b the impulse in the string when it jerks tight, in terms of m.

Step 1: Draw a diagram for the initial and final situations.

a Before collision: 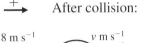 After collision:

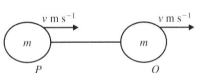

Note:
Only one particle has a velocity before the jerk but they both have the same velocity after the jerk.

Step 2: Calculate the total momentum before and after the collision.

Momentum before jerk	Momentum after jerk
$= 8m$	$= mv + mv$
	$= 2mv$

Note:
Let motion from left to right be positive.

Step 3: Calculate unknowns using the principle of conservation of momentum.

By the principle of conservation of momentum:
$$8m = 2mv$$
$$v = 4$$

So, the combined speed after the string becomes taut is 4 m s^{-1}.

b The magnitude of the impulse in the string will be equal to the magnitude of the change in momentum of either particle. For particle P:

$$I = 4m - 0m = 4m$$

So, the magnitude of the impulse in the string is $4m$ N s.

Recall:
$\mathbf{I} = m\mathbf{v} - m\mathbf{u}$

Note:
The result is the same using particle Q.

SKILLS CHECK **5C: Momentum and impulse**

1 Calculate the momentum for a particle of mass m kg and travelling with velocity v m s^{-1} when:

a $m = 2$, $v = 4$ **b** $m = 0.1$, $v = 18$ **c** $m = \frac{2}{3}$, $v = 15$.

2 Find the magnitude of the change in momentum of a particle of mass 8 kg that changes its speed from 2 m s⁻¹ to:

   a 14 m s⁻¹ in the same direction   **b** 14 m s⁻¹ in the opposite direction.

3 A particle of mass 2 kg is travelling in a horizontal line with speed 3 m s⁻¹ when a force of 15 N acts on it in the same direction for 4 seconds.

   a Find the impulse that the force exerts on the particle.

   b Find the final velocity of the particle, after 4 seconds.

 4 A particle of mass 10 kg travels across a smooth horizontal table. The particle increases its speed from 80 m s⁻¹ to 100 m s⁻¹ under the action of a constant force F N, which acts for 0.5 seconds. Find F.

5 At $t = 0$ a particle of mass 3 kg is travelling with velocity $(7\mathbf{i} - 2\mathbf{j})$ m s⁻¹ where t is the time measured in seconds. At $t = 4$ it has velocity $(2\mathbf{i} - 7\mathbf{j})$ m s⁻¹. Find the magnitude of the constant force that acts on the particle.

 6 A particle of mass m kg travels in a straight horizontal line towards a wall with speed u m s⁻¹. After impact with the wall the speed of the ball is halved. Find u if the wall exerts an impulse of $(15m)$ N s on the ball.

7 In parts **a**, **b** and **c**, two particles collide as shown. The diagrams show the situation before and after the impact. Find the unknown initial velocity u or the unknown final velocity v as appropriate:

   a Before collision: After collision:

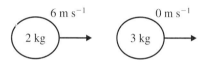 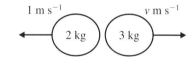

   b Before collision: After collision:

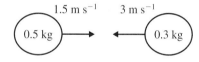 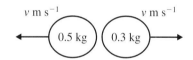

   c Before collision: After collision:

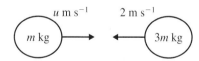

8 A car of mass 800 kg travelling on a smooth road with speed 6 m s⁻¹ collides and coalesces with a stationary car of mass 500 kg. Find the speed of the combined cars after the collision. Also find the magnitude of the impulse exerted by the moving car on the stationary one.

 9 A gun of mass 3 kg fires a bullet of mass 20 g. If the gun recoils at 2 m s⁻¹, find the speed of the bullet after the shot is fired. The bullet drives into a wall with this speed and stops after it has driven 2 cm into the wall in a horizontal direction. Find the magnitude of the resistance provided by the wall.

10 Two particles A and B of mass M kg and $2M$ kg respectively are connected by a light, inextensible string, which is initially slack. Particle B is projected away from particle A with speed u m s⁻¹. Find u if the combined speed of the two particles after the string becomes taut is 4 m s⁻¹. Find the impulse in the string, in terms of M, when the string jerks tight.

SKILLS CHECK **5C EXTRA is on the CD**

1 A small stone moves horizontally in a straight line across the surface of an ice rink. The stone is given an initial speed of 7 m s⁻¹. It comes to rest after moving a distance of 10 m. Find

a the deceleration of the stone while it is moving,

b the coefficient of friction between the stone and the ice. [Edexcel June 2000]

 2

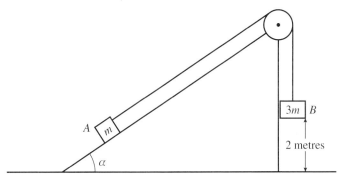

Two particles A and B have masses $3m$ and km respectively, where $k > 3$. They are connected by a light inextensible string which passes over a smooth fixed pulley. The system is released from rest with the string taut and the hanging parts of the string vertical, as shown in the diagram. While the particles are moving freely, A has an acceleration of magnitude $\frac{2}{5}g$.

a Find, in terms of m and g, the tension in the string.

b State why B also has an acceleration of magnitude $\frac{2}{5}g$.

c Find the value of k.

d State how you have used the fact that the string is light. [Edexcel Jan 2001]

3

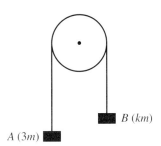

A particle A, of mass m, lies on a rough inclined plane of angle α, where $\tan \alpha = \frac{3}{4}$. The coefficient of friction between A and the plane is $\frac{1}{4}$. The particle A is connected to a particle B, of mass $2m$, by a light inextensible string. The string lies along a line of greatest slope of the plane and passes over a small smooth pulley fixed at the top of the plane. The system is held at rest with the string taut and with B hanging 2 metres above a horizontal plane, as shown above.

The system is now released from rest.

a Find the initial acceleration of A.

b State clearly where the facts that
   i the string is "inextensible",   **ii** the pulley is "smooth"

have been used in your calculation for part **a**.

After A has travelled a distance of 1 metre up the plane, the string breaks. Assuming that A does not hit the pulley during the subsequent motion,

c calculate the further distance up the plane that A travels before first coming to instantaneous rest.

[Edexcel Jan 1997]

4

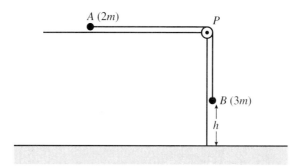

A particle A of mass $2m$ moves on the surface of a rough horizontal table and is attached to one end of a light inextensible string. The coefficient of friction between A and the table is $\frac{1}{3}$. The string passes over a small smooth pulley P fixed at the edge of the table. The other end of the string is attached to a particle B of mass $3m$ which hangs freely below the pulley, as shown in the diagram above. The system is released from rest with the string taut and with B at a height h above the ground.

a Find the tension in the string while the particles are moving.

Given that A does not reach P before B reaches the ground,

b show that the speed with which B reaches the ground is $\sqrt{\left(\dfrac{14gh}{15}\right)}$.

When B reaches the ground, it does not rebound and A continues to move along the table. Given that A does not reach P before coming to rest,

c show that the total length of the string must be at least $\dfrac{12h}{5}$.   [Edexcel June 1999]

5 Two small balls A and B have masses 0.5 kg and 0.2 kg respectively. They are moving towards each other in opposite directions on a smooth horizontal table when they collide directly. Immediately before the collision, the speed of A is $3\,\text{m s}^{-1}$ and the speed of B is $2\,\text{m s}^{-1}$. The speed of A immediately after the collision is $1.5\,\text{m s}^{-1}$. The direction of motion of A is unchanged as a result of the collision.

By modelling the balls as particles, find

a the speed of B immediately after the collision,

b the magnitude of the impulse exerted on B in the collision.   [Edexcel June 2001]

6 A truck A of mass 6000 kg is moving with a speed of $12\,\text{m s}^{-1}$ along a straight horizontal railway line when it collides with another truck B of mass 9000 kg which is stationary. After the collision the two trucks move on together.

a Find the speed of the trucks immediately after the collision.

b Find the magnitude of the impulse exerted on B when the trucks collide, stating the units in which your answer is given.

After the collision, the motion of the two trucks is opposed by a constant horizontal resistance of magnitude R Newtons. The trucks come to rest 20 s after the collision.

c Find R.   [Edexcel June 1998]

6 Moments

6.1 Moment of a force and equilibrium with parallel forces

Moment of a force.

Suppose a rod is fixed at one end O, but is free to rotate around that end. If a force of F N is then applied to the rod at a distance d m from the pivot at O, there is a turning effect due to the force.

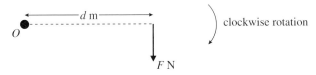

The greater the distance from the fixed point, the greater the turning effect of the force. You can measure the turning effect: it is called the **moment** of the force about a fixed point and has units of Nm.

Moment = force × perpendicular distance from the fixed point to the line of action of F

i.e.   Moment = Fd

Note:
The turning effect can be **clockwise** or **anticlockwise**. This is called the **sense of rotation** and it is important to state this when describing the turning effect.

Example 6.1 Find the moment of the force about the point O, stating the sense of rotation.

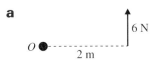

b O •- - - - - -→ 6 N
   2 m

Step 1: Calculate the moment of the force.

a Moment = Fd
$$= 6 \times 2$$
$$= 12$$

The moment is 12 N m anticlockwise.

b The distance between the line of action of the force and point O is 0 m, so the moment of the force about O is 0 N m.

Overall moment of more than one force

When more than one force acts about a point of rotation, the overall moment is the sum of the component moments, where moments with the same sense of rotation are added whilst those with an opposing sense of rotation are subtracted. These problems can be solved using a moments table, as shown in the next example.

Example 6.2 The diagram shows three forces that are applied to a light rod PQ of length 10 m. Find the overall moment about the point P.

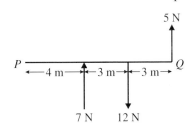

Tip:
Use the abbreviations cw for clockwise and acw for anticlockwise.

Recall:
Moment = Fd

Moments about P			
Force (N)	7	12	5
Distance from P (m)	4	7	10
Sense and moment (N m)	+28	−84	+50

Note:
The convention used is + for acw and − for cw.
Sum of moments
$= (28 − 84 + 50)$ N m
$= −6$ N m.

Total clockwise moment $= 84$ N m

Total anticlockwise moment $= (28 + 50)$ N m $= 78$ N m

Step 2: Find the difference between the total cw and acw moments.

Overall moment $= (84 − 78)$ N m
$\qquad\qquad\quad = 6$ N m clockwise

Equilibrium of coplanar forces acting on a body

A system of forces acting on a body is in equilibrium when there is no resultant force and no overall turning effect. Only moments of parallel forces acting on a body in equilibrium are considered in M1. The following two conditions hold when a system of parallel forces acting on a body is in equilibrium.

1 **The resultant force in any direction is 0.**

2 **About any point, total clockwise moments = total anticlockwise moments.**

(**or the sum of moments about any point is 0**)

If the body is not light then it has a weight, which will also have a moment about a point. If the rod is uniform then the weight will act in the middle of the rod. This point is called the **centre of mass** of the body.

Example 6.3   A uniform beam AB of mass 6 kg and length 4 m rests on a pivot at the point C where $AC = 2.5$ m. A block of mass m kg is placed on the rod at the point B so that the rod is in equilibrium. Find the mass of the block and the magnitude of the reaction at the support.

Note:
Let the reaction force at the pivot be R N.

Step 1: Draw a diagram with all the distances and forces included.

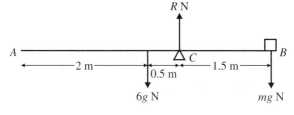

Note:
There is always a reaction force exerted at a pivot.

Recall:
If the mass is m kg then the weight is mg N.

Step 2: Take moments about any point using the moments table (sum of moments = 0 if in equilibrium).

Moments about C		
Force (N)	mg	$6g$
Distance from C (m)	1.5	0.5
Sense and moment (N m)	$−1.5\,mg$	$+3\,g$

Tip:
Take moments about a point where a force is not known; this force will not contribute to the calculation.

In equilibrium:

Method 1

Sum of moments $= 0$

$$-1.5\,mg + 3\,g = 0 \qquad\qquad \dots (1)$$

Method 2

Sum of cw moments $=$ sum of acw moments

$$1.5\,mg = 3\,g \qquad\qquad \dots (1)$$

Note:
Both methods lead to the same solution.

Step 3: Resolve forces vertically (and equate to 0 if in equilibrium).	Resolving vertically:

$$R - 6g - mg = 0 \qquad \text{... (2)}$$

Step 4: Solve for the unknowns.

$$m = 2 \qquad \text{... from (1)}$$

The mass of the block is 2 kg.

$$R = 6g + mg \qquad \text{...from (2)}$$
$$= 78.4$$

The magnitude of the reaction at the support is 78.4 N.

Tip:
The force of the support on the plank is another way of referring to the reaction force.

Example 6.4 A uniform rod PQ of mass 15 kg and length 10 m rests on two supports at R and S where $PR = 2$ m and $SQ = 3$ m. Zaheer has mass 20 kg and sits at P. Nadia has mass 10 kg and sits at the point T where $PT = 8$ m. Find the magnitude of the force of the support on the plank at each of the supports.

Note:
Let the reaction forces at R and S be X N and Y N respectively.

Step 1: Draw a diagram with all the distances and forces included.

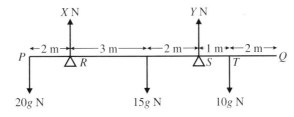

Step 2: Take moments about any point using the moments table (sum of moments = 0 if in equilibrium).

Tip:
You could take moments about either R or S as this will also eliminate one of the unknowns.

Moments about R ($R\curvearrowright$)				
Force (N)	$20g$	$15g$	Y	$10g$
Distance from R (m)	2	3	5	6
Sense and moment (N m)	$+40g$	$-45g$	$+5Y$	$-60g$

In equilibrium: $40g + 5Y - 45g - 60g = 0$ ... (1)

Step 3: Resolve forces vertically (and equate to 0 if in equilibrium).

Resolving vertically: $X + Y - 20g - 15g - 10g = 0$ ... (2)

Note:
$R\curvearrowright$ is sometimes used as a short form for take moments about R.

Step 4: Solve for the unknowns.

$$Y = 127.4 \qquad \text{... from (1)}$$
$$X = 20g + 15g + 10g - Y = 313.6 \qquad \text{... from (2)}$$

The magnitude of the force of the support on the plank at R is 314 N.

The magnitude of the force of the support on the plank at S is 127 N.

If a rod is not uniform then the centre of mass may not be at the centre of the rod. It may act anywhere along the rod.

Example 6.5 Two sand bags of masses 7 kg and 3 kg are placed on the ends of a non-uniform rod PQ of mass 8 kg and length 4 m, with the 7 kg mass placed at P. The rod rests in equilibrium on the edge of a table, with half of the rod lying on the table's surface. Find the distance of the centre of mass from the edge of the table and the reaction force of the table on the rod.

Note:
Let the distance of the centre of mass from the centre of the rod be x m and the reaction force be R N.

Step 1: Draw a diagram with all the distances and forces included.

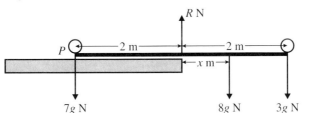

Note:
The rod will pivot about the edge of the table and so there will be a reaction force at the pivot here.

Moments about centre of rod			
Force (N)	7g	3g	8g
Distance from centre of rod (m)	2	2	x
Sense and moment (N m)	+14g	−6g	−8gx

In equilibrium: $+14g - 8gx - 6g = 0$ … (1)

Step 3: Resolve forces vertically (and equate to 0 if in equilibrium).

Resolving vertically:

$$R - 7g - 8g - 3g = 0 \qquad\qquad \text{… (2)}$$

Step 4: Solve for the unknowns.

$$x = 1 \qquad\qquad \text{… from (1)}$$

$$R = 18g \qquad\qquad \text{… from (2)}$$

$$= 176.4\ldots$$

Note:
Sometimes you may be asked to leave your answer in terms of g.

The distance from the centre of mass of the rod to the edge of the table is 1 m and the reaction force of the table on the rod is 180 N (2 s.f.).

Example 6.6 A non-uniform plank *AB* of mass 15 kg and length 8 m rests on two pivots at *C* and *D* where *AC* = *DB* = 2 m. The centre of mass of the plank is 3 m from *A*. David has mass 25 kg and sits at *A*. Shehzma has the same mass and sits somewhere along the plank to maintain equilibrium. Find where Shehzma sits, given that the magnitude of the reaction force at *C* is twice the magnitude of the reaction force at *D*. Also, find the magnitude of the reaction forces at each of the pivots.

Step 1: Draw a diagram with all the distances and forces included.

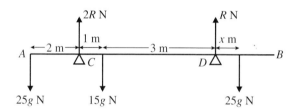

Note:
Let the reaction forces at *D* be *R* N. Then the reaction force at *C* is 2*R* N. Let the distance of the child from *D* be *x* m.

Step 2: Take moments about any point using the moments table (sum of moments = 0 if in equilibrium).

Moments about C				
Force (N)	25g	15g	R	25g
Distance from P (m)	2	1	4	(4 + x)
Sense and moment (N m)	+50g	−15g	+4R	−25g(4 + x)

In equilibrium: $50g - 15g + 4R - 25g(4 + x) = 0$ … (1)

Step 3: Resolve forces vertically (and equate to 0 if in equilibrium).

Resolving vertically:

$$2R + R - 25g - 15g - 25g = 0 \qquad\qquad \text{… (2)}$$

Step 4: Solve for the unknowns.

$$3R = 65g \qquad\qquad \text{… from (2)}$$

$$R = 212.333\ldots$$

$$25g(4 + x) = 50g + 4R - 15g \quad \text{… from (1)}$$

$$x = 0.866\ldots$$

The distance that Shehzma must sit in order to maintain equilibrium is 0.87 m from the support at *D*. The magnitude of the reaction force at *C* is *R* N = 212 N (3 s.f.) and at *D* is 2 *R* N = 425 N (3 s.f.).

Tilting

When a beam is placed on two supports and a force, F N, is applied to either end of the beam, then if F is large enough the beam can tilt about one of the pivots:

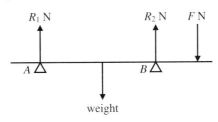

weight

Note:
When the rod is about to rotate about pivot B, it is about to lose contact with pivot A, hence $R_1 = 0$.

At the point of tilting about B the reaction force at A will be 0 N, while the situation is still in equilibrium.

Example 6.7 A uniform beam AB of mass 12 kg and length 6 m rests on two pivots at P and Q, where $AP = 1$ m and $QB = 1.5$ m. A particle of M kg is placed at A and the beam is about to tilt about the pivot at P. Find the mass of the particle and the reaction force at P.

Step 1: Draw a diagram with all the distances and forces included.

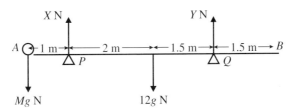

Mg N $12g$ N

Note:
Let the reactions at the supports P and Q be X N and Y N respectively.

Step 2: Take moments about any point using the moments table (sum of moments = 0 if in equilibrium).

Moments about P		
Force (N)	Mg	$12g$
Distance from P (m)	1	2
Sense and moment (N m)	$+Mg$	$-24g$

Tip:
Take moments about P to eliminate the unknown reaction force X, and as the beam tilts about the pivot at P then the reaction force Y is 0 at Q.

In equilibrium: $Mg - 24g = 0$ ... (1)

Step 3: Resolve forces vertically (and equate to 0 if in equilibrium).

Resolving vertically:

$$X - Mg - 12g = 0 \qquad\qquad\qquad \text{... (2)}$$

Step 4: Solve for the unknowns.

$$M = 24 \qquad\qquad \text{... from (1)}$$

$$X = Mg + 12g = 36g \qquad \text{... from (2)}$$

$$= 352.8$$

The mass of the particle placed at A is 24 kg and the reaction force at P is 353 N (3 s.f.).

SKILLS CHECK   **6A: Moment of a force and equilibrium with parallel forces**

1 For parts **a**, **b**, and **c** find the overall moment of the force about O, stating also the sense of rotation.

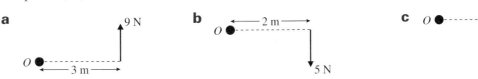

2 Forces are applied to a light rod PQ at distances shown in the diagrams. Find the overall moment of the forces about the point O, stating the sense of the rotation.

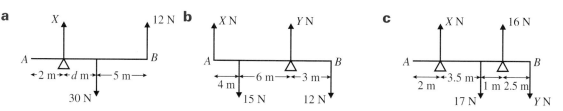

3 In parts **a**, **b**, and **c** find the forces X or Y or the distance d as shown when the pivoted light rod AB is in equilibrium.

a

X 12 N

A ———△——————— B
  ←2 m→←d m→←—5 m—→
      30 N

b

X N Y N

A ————————△——— B
  ←—6 m—→←3 m→
 4 m
    15 N   12 N

c

X N   16 N

A —△——————△— B
  ←—3.5 m—→←1 m 2.5 m
 2 m
    17 N Y N

4 A uniform rod AB of length 4 m is pivoted at the point C where $AC = 1.5$ m. A particle of mass 30 kg is attached at A on the rod, which keeps the rod in equilibrium. Find the reaction force at the pivot and the mass of the rod.

5 Two children of mass 30 kg and 10 kg are sitting on the ends P and Q respectively of a uniform seesaw. The seesaw has mass M kg and length 6 m, and it is pivoted at the point R where $PR = 2$ m. Find M and explain the significance of modelling the children as particles.

6 A non-uniform beam AB of length 8 m and mass 2 kg rests over the edge of the table, with 5 m of the beam on the table and 3 m hanging over the edge freely. The centre of mass of the beam is 1 m from B. A particle of mass 12 kg is placed on the beam to maintain equilibrium. Find the distance of the particle from A and the magnitude of the force of the table on the plank.

7 A non-uniform plank PQ of length 10 m and mass 26 kg is pivoted at the points R and S, where PR is 1 m and PS is 8 m. A boy of mass 22 kg stands at a point 3 m from Q to maintain equilibrium. Given that the magnitude of the force of the pivot on the plank at R is twice the magnitude of the force of the pivot on the plank at S, find the distance of the centre of mass of the plank from P and the reaction forces at each of the pivots.

A load is placed at the point Q and the plank is now on the point of turning about S. Find the mass of the load.

8 A non-uniform plank AB has length $7x$ m and mass M kg. It rests on two supports C and D where $AC = x$ m and $CD = 3x$ m. A block of mass $2M$ kg is placed at A and the rod is about to tilt about C. Calculate, in terms of x, the distance of the centre of mass from A.

SKILLS CHECK **6A EXTRA** is on the CD

Examination practice   Moments

1

A uniform rod AB has length 8 m and mass 12 kg. A particle of mass 8 kg is attached to the rod at B. The rod is supported at a point C and is in equilibrium in a horizontal position, as shown above.

Find the length of AC.

[Edexcel Jan 1997]

2

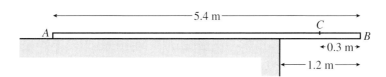

A plank of wood *AB* has length 5.4 m. It lies on a horizontal platform, with 1.2 m projecting over the edge, as shown above. When a girl of mass 50 kg stands at the point *C* on the plank, where *BC* = 0.3 m, the plank is on the point of tilting. By modelling the plank as a uniform rod and the girl as a particle,

a find the mass of the plank.

The girl places a rock on the end of the plank at *A*. By modelling the rock also as a particle,

b find, to 2 significant figures, the smallest mass of the rock which will enable the girl to stand on the plank at *B* without it tilting.

c State briefly how you have used the modelling assumptions that
  i the plank is uniform,
  ii the rock is a particle. [Edexcel June 1999]

3

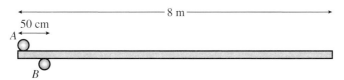

A large uniform plank of wood of length 8 m and mass 30 kg is held in equilibrium by two small steel rollers *A* and *B*, ready to be pushed into a saw-mill. The centres of the rollers are 50 cm apart. One end of the plank presses against roller *A* from underneath, and the plank rests on top of roller *B*, as shown above. The rollers are adjusted so that the plank remains horizontal and the force exerted on the plank by each roller is vertical.

a Suggest a suitable model for the plank to determine the forces exerted by the rollers.

b Find the magnitude of the force exerted on the plank by the roller at *B*.

c Find the magnitude of the force exerted on the plank by the roller at *A*. [Edexcel Jan 1996]

4

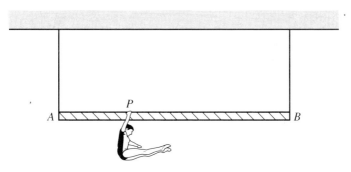

A gymnast of mass 35 kg hangs by one hand from the point *P* on a bar *AB* of length 3 m and mass 12 kg. The bar is suspended by two vertical cables which are attached to the ends *A* and *B*, and it is hanging in equilibrium in a horizontal position, as shown above. The tension in the cable at *A* is twice the tension in the cable at *B*. By modelling the bar as a uniform rod, and the gymnast as a particle,

a find the distance *AP*.

b State two ways in which, in your calculation, you have used the model of the bar as a "uniform rod".

[Edexcel June 1997]

5

A large log AB is 6 m long. It rests in a horizontal position on two smooth supports C and D, where $AC = 1$ m and $BD = 1$ m, as shown. David needs an estimate of the weight of the log, but the log is too heavy to lift off both supports. When David applies a force of magnitude 1500 N vertically upwards to the log at A, the log is about to tilt about D.

a State the value of the reaction on the log at C for this case.

David initially models the log as a uniform rod. Using this model,

b estimate the weight of the log.

The shape of the log convinces David that his initial modelling assumption is too simple. He removes the force at A and applies a force acting vertically upwards at B. He finds that the log is about to tilt about C when this force has magnitude 1000 N. David now models the log as a non-uniform rod, with the distance of the centre of mass of the log from C as x metres. Using this model, find

c a new estimate for the weight of the log.

d the value of x.

e State how you have used the modelling assumption that the log is a rod.   [Edexcel June 2001]

Practice exam paper

Answer **all** questions.

Time allowed: 1 hour 30 minutes

A calculator **may** be used in this paper.

1 A constant force **F** acts on a particle P of mass 4 kg for 3 seconds causing its velocity to change. The initial velocity of P is $(-4\mathbf{i} + 5\mathbf{j})$ m s^{-1} and after the 3 seconds its velocity is $(14\mathbf{i} - 19\mathbf{j})$ m s^{-1}.

   a Find the acceleration of P.   *(2 marks)*

   b Find the magnitude of **F**.   *(3 marks)*

   c Find, to the nearest degree, the acute angle between the line of action of **F** and the vector **i**.   *(2 marks)*

2 A particle P of mass $4m$ is moving along a straight line with constant speed $2u$. It collides with another particle Q of mass $3m$ which is moving with the same speed along the same line but in the opposite direction. As a result of the collision P is brought to rest.

   a Find the speed of Q after the collision and state its direction of motion.   *(4 marks)*

   b Find the magnitude of the impulse, I, exerted by P on Q in the collision.   *(3 marks)*

3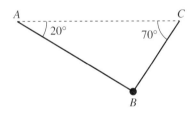

A particle of mass 0.7 kg is attached to the end B of two ropes AB and BC and hangs in equilibrium. The rope AB makes an angle of 20° with the horizontal and the rope BC makes an angle of 70° with the horizontal, as shown in the diagram. The ropes are modelled as light and inextensible strings.

   a Find the tension in each of the strings AB and BC.   *(7 marks)*

   b Explain how you have used the assumption that the ropes are light.   *(1 mark)*

4 A uniform beam AB has mass 20 kg and length 4 m. The beam is suspended by two vertical ropes attached to the beam, one at A and the other at C, where $BC = 1$ m. A particle of mass 12 kg is placed on the beam, 50 cm from B, and another particle of mass 10 kg is placed on the beam between A and B. The beam is in equilibrium in a horizontal position, and the tension in the rope attached at C is five times the tension in the rope attached at A.

   a Find the tension in the rope attached at C.   *(3 marks)*

   b Find the distance between the particles.   *(5 marks)*

5 In a simple model of the motion of a car, its velocity **v** m s^{-1}, at time t seconds, is given by
$$\mathbf{v} = (3t^2 - 4t + 1)\mathbf{i} + (5t - 3)\mathbf{j} \quad \text{where} \quad t \geqslant 0.$$

   a Find the initial speed of the car.   *(3 marks)*

   b Find when the car is moving parallel to the vector $(\mathbf{i} - \mathbf{j})$.   *(5 marks)*

6

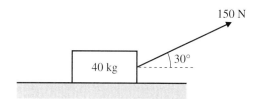

A large crate of mass 40 kg lies at rest on rough horizontal ground. One end of a rope is attached to the crate and the rope makes an angle of 30° with the ground as shown in the diagram. The tension in the rope is 150 N and the crate is on the point of moving along the plane.

By modelling the crate as a particle,

a find the value of μ, the coefficient of friction between the crate and the ground. *(5 marks)*

The rope remains at 30° to the ground but the tension in it is increased to 200 N.

b Find the acceleration of the crate. *(5 marks)*

7 Two motorcycles P and Q move in the same direction along a straight horizontal road. At time $t = 0$, P sets off with speed 1 m s^{-1} from the point O and moves with constant acceleration 2 m s^{-2}. Four seconds later Q sets off in pursuit from O with initial speed 16 m s^{-1} and moves with constant acceleration 1 m s^{-2}. Find

a the times between which motorcycle Q is in front of motorcycle P, *(9 marks)*

b the distance from O at which motorcycle P overtakes motorcycle Q. *(3 marks)*

8

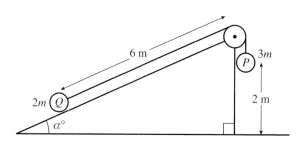

A particle P of mass $3m$ kg is attached to one end of a light inextensible string of length 6 m. The other end of the string is attached to another particle Q of mass $2m$ kg which is held at rest on a rough plane inclined to the horizontal at an angle α where $\tan \alpha° = \frac{5}{12}$. The string lies along a line of greatest slope of the plane and passes over a small smooth pulley which is fixed at the top of the plane. The particle P hangs freely just below the pulley, a distance of 2 m above horizontal ground, as shown in the diagram. The coefficient of friction between Q and the plane is $\frac{11}{36}$. The system is released from rest.

a Find, in terms of g, the acceleration of P before it hits the ground. *(8 marks)*

b Find the distance of Q from the pulley when it first comes to instantaneous rest. *(7 marks)*

Answers

SKILLS CHECK 2A (page 8)

1   **a**   5 km, 307°   **b**   5.73 km, 282°   **c**   4.53 km, 314°
2   **a** $\mathbf{a} + \mathbf{b}$ **b** $\mathbf{a}$ **c** $\frac{1}{2}\mathbf{a}$
     d $\frac{1}{2}\mathbf{a} + \mathbf{b}$ **e** $-\mathbf{a} - \frac{1}{2}\mathbf{b}$ **f** $\frac{1}{2}(\mathbf{b} - \mathbf{a})$
3   **a** $\mathbf{b} - \mathbf{a}$ **b** $\mathbf{a} - \mathbf{b}$
     c $\frac{3}{4}(\mathbf{b} - \mathbf{a})$ **d** $\frac{1}{4}(\mathbf{a} + 3\mathbf{b})$
4   **a**   5, 53.1°   **b**   6.08, 351° (or $-9.46°$)
     c   2.83, 225°   **d**   6.71, 117°
5   **a** $8.46\mathbf{i} + 3.08\mathbf{j}$ **b** $-3.94\mathbf{i} + 0.69\mathbf{j}$ **c** $-12.1\mathbf{i} - 7\mathbf{j}$
6   **a** $5\mathbf{i} - 3\mathbf{j}$, 5.83, 121°   **b** $\mathbf{i} + 5\mathbf{j}$, 5.10, 11.3°   **c** $8\mathbf{i} - 2\mathbf{j}$, 8.25, 104°
     d $-5\mathbf{i} - 11\mathbf{j}$, 12.1, 156°   **e** $4\mathbf{i} + 6\mathbf{j}$, 7.21, 33.7°
7   **a**   4   **b** $-\frac{33}{4}$ **c** ± 12
8   **a** $\frac{3}{5}\mathbf{i} + \frac{4}{5}\mathbf{j}$ **b** $\dfrac{4}{\sqrt{80}}\mathbf{i} - \dfrac{8}{\sqrt{80}}\mathbf{j}$
9   **a** $\frac{7}{25}\mathbf{i} + \frac{24}{25}\mathbf{j}$ **b** $14\mathbf{i} + 48\mathbf{j}$ **c** $21\mathbf{i} + 72\mathbf{j}$
10   **a** $\mu = \frac{11}{24}$ **b** $\mu = -\frac{3}{2}$

SKILLS CHECK 2B (page 13)

1   **a** $\mathbf{i} + \mathbf{j}$, 1.41 m s^{-1} **b** $-\mathbf{i} + 2\mathbf{j}$, 2.24 m s^{-1} **c** $1.5\mathbf{i} + 5\mathbf{j}$, 5.22 m s^{-1}
2   **a** $(2\mathbf{i} + 24\mathbf{j})$ m   **b** $(24\mathbf{i} + 25\mathbf{j})$ m   **c**   34.7 m, 46.2°
3   **a** $(1.5\mathbf{i} + 4.5\mathbf{j})$ m s^{-2} **b** $(16\mathbf{i} - 16\mathbf{j})$ m s^{-2} **c** $\mathbf{j}$ m s^{-2}
4   20 m s^{-1}
5   **a** $[(3\mathbf{i} + 8\mathbf{j}) + t(\mathbf{i} - 2\mathbf{j})]$ m $= [(3 + t)\mathbf{i} + (8 - 2t)\mathbf{j}]$ m
     b $(6\mathbf{i} + 2\mathbf{j})$ m   **c**   2
6   **a** $(3\mathbf{i} + 14\mathbf{j})$ m, 14.3 m   **b** $6\mathbf{i} + 8\mathbf{j}$ m s^{-1}; $(-2\mathbf{i} - \mathbf{j})$ m s^{-1}
     c $(7\mathbf{i} + 4\mathbf{j})$ m
7   **a** $(t\mathbf{i} + 3t\mathbf{j})$ km   **b** $[(10 - t)\mathbf{i} + (10 + 2t)\mathbf{j}]$ km
     c   10 km   **e**   18:00, 4.47 km
8   He does meet the ball, at $(-4\mathbf{i} - 2\mathbf{j})$ m. Assumed no air resistance for ball (it is small) and for cricketer (as his speed is not fast), treated ball and man as particle.
9   **a**   16, 1, 1 m

Exam Practice 2 (page 15)

1   **a**   5.83 m s^{-1} **b**   9.43 m
2   **a** $(2\mathbf{i} + \mathbf{j})$ m s^{-1} **b**   26.6°   **c**   12.6 m
3   **a** $v_C = 20\mathbf{j}$, $v_D = -10\mathbf{i}$
     b $r_C = -50\mathbf{i} + 20t\mathbf{j}$, $r_D = -10t\mathbf{i}$
     d   Destroyer not detected
4   **a** $8\mathbf{i} + 6\mathbf{j}$ **c**   23.4 m
5   **a** $10\sqrt{3}\mathbf{i} + 10\mathbf{j}$ **b** $r = 30t\mathbf{j}$ $s = 10\sqrt{3}t\mathbf{i}$ **c**   1400 h

SKILLS CHECK 3A (page 19)

1   **a**   2 m s^{-2} **b**   4 m
2    $\frac{25}{18}$ m s^{-2}
3   9 m s^{-1}
4   **a** $\frac{2}{3}$ s   **b**   No air resistance, car is a particle
5   50 m
6   **a**   13.9 s   **b**   965 m
7   **a**   2.5 m s^{-2} **b**   140 m
8   **a**   4 m   **b**   4.674 s   **c**   10.8 m

SKILLS CHECK 3B (page 26)

1   8.4 m s^{-1}, 36.4 m
2   40 m, 5.7 s, 4.04 s
3   4.04 s, book is a particle, no air resistance, gravity constant, book starts from rest
4   **a**   14 m   **b**   3.12 s
5   31.3 m s^{-1}
6   **a**   5.63 s   **b**   155.3 m
7   **b**   6.38 m

8   **a**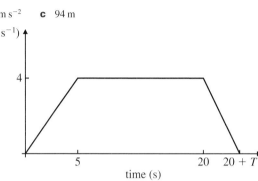
     b   2 m s^{-2}, $-\frac{4}{5}$ m s^{-2} **c**   94 m

9   **a**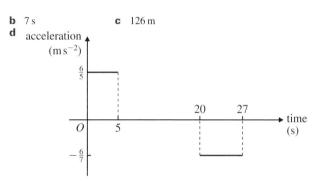
     b   7 s   **c**   126 m
     d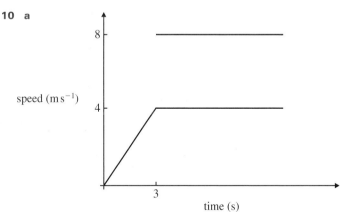

10   **a**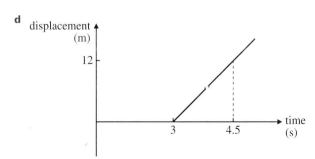
     b   4.5 s   **c**   12 m
     d

Exam Practice 3 (page 27)

1 a $38.3\,\text{m s}^{-1}$ **b** $3.91\,\text{s}$ **c** Air resistance, wind
2 a v **b** $120\,\text{s}$ **c** $3780\,\text{m}$

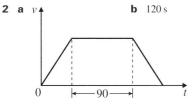

3 a $12\,\text{m s}^{-2}$ **b** $80\,\text{m s}^{-1}$
4 a v **b** $9.4\,\text{m s}^{-1}$

5 b v **c** $9\,\text{s}$ **d** $72\,\text{m}$

SKILLS CHECK 4A (page 33)

1 a $6.7\,\text{N}, 63.4°$ **b** $7.1\,\text{N}, 135.0°$ **c** $20.6\,\text{N}, 29.05°$
   d $3.6\,\text{N}, 33.7°$ **e** $5.8\,\text{N}, 301.0°$ (or $-59.0°$)
2 a $9.8\,\text{N}, 1.7\,\text{N}$ **b** $24.1\,\text{N}, 6.5\,\text{N}$ **c** $3.1\,\text{N}, 34.9\,\text{N}$
3 a $-2.5\mathbf{i} + 4.3\mathbf{j}$ **b** $17.3\mathbf{i} - 10.0\mathbf{j}$ **c** $-111.1\mathbf{i} - 93.2\mathbf{j}$
4 a $16.2\,\text{N}, 4.0°$ **b** $98.5\,\text{N}, 8.54°$
5 a $15.1\,\text{N}, 7.59°$ **b** $5.39\,\text{N}, 248°$
6 a $10\,\text{N}, 8.45\,\text{N}$ **b** $18.8\,\text{N}, 12.3\,\text{N}$
7 $11.7\,\text{N}, 190°$
8 a $-14, -4$ **b** $1, 10$ **c** $2, 14$

SKILLS CHECK 4B (page 40)

1 a $1.96\,\text{N}, 7\,\text{N}$ **b** $29.4\,\text{N}, 14\,\text{N}$ **c** $88.2\,\text{N}, 24\,\text{N}$
2 a $10.4\,\text{N}, 43\,\text{N}$ **b** $13.9\,\text{N}, 0.91\,\text{kg}$ **c** $6.93\,\text{N}, 5.36\,\text{N}$
3 a $150\,\text{N}$ **b** 0.01
4 a $24.8\,\text{N}, 0.78$ **b** $1.36\,\text{N}, 5.25\,\text{N}$ **c** $0.97, 1.95\,\text{kg}$
5 $73.6\,\text{N}, 69.1\,\text{N}$
6 $12.3\,\text{N}, 1.61\,\text{kg}$
7 $\frac{12}{13}mg, \frac{5}{13}mg, \frac{5}{12}$
8 a $27.4\,\text{N}$ **b** $1.97\,\text{N}$
9 a $44.5\,\text{N}$ **b** $5.57\,\text{N}$
10 $5180\,\text{N}, 13\,300\,\text{N}$

Exam Practice 4 (page 41)

1 a $17.3\,\text{N}$ **b** $21.3\,\text{N}$
2 a   **b** $\frac{2}{11}mg$ **c** F acts downwards.

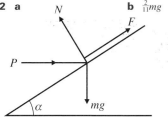

3 $363\,\text{N}$
4 a $53°$ **b** $25\,\text{N}$
5 a $-6\mathbf{i} + 9\mathbf{j}$ **b** $10.8\,\text{N}$ **c** $33.7°$

SKILLS CHECK 5A (page 46)

1 a $4\,\text{m s}^{-2}$ **b** $2\,\text{m s}^{-2}$ **c** $5\,\text{m s}^{-2}$ **d** $0.56\,\text{m s}^{-2}$
2 a $6\,\text{N}$ **b** $1.65\,\text{N}$ **c** $0.27\,\text{N}$
3 $2.35\,\text{m s}^{-2}, 18.8\,\text{m}$
4 $4500\,\text{N}$
5 $4273\,\text{N}$
6 $39.9\,\text{kg}$. Wall treated as a particle, neglect air resistance, rope is light, inextensible.
7 $6.9\,\text{m s}^{-2}$
8 $8780\,\text{N}$
9 $\frac{5}{12}, 1.62\,\text{m s}^{-2}, 3.23\,\text{m}$
10 $3.16\,\text{m s}^{-2}, -34.7°$
11 $3.5, -2$

SKILLS CHECK 5B (page 52)

1 $15\,\text{N}, 10\,\text{N}, 16\,\text{N}$
2 $2.7\,\text{N}, 0.9\,\text{N}, 0.13\,\text{m s}^{-2}$; the rope is light and inextensible, and there is no air resistance.
3 $7.2\,\text{m s}^{-2}, 34\,\text{N}$
4 $6240\,\text{N}, 14040\,\text{N}$; the chain is light and the rope is light and inextensible.
5 $2.45\,\text{m s}^{-2}, 1.56\,\text{s}, 73.5\,\text{N}$
6 $3.27\,\text{m s}^{-2}, 3.61\,\text{m s}^{-1}, 4.67\,\text{m}$; tensions are the same in the string either side of the pulley; there is no component weight for the string; there is no force used in extending the string.
7 0.803
8 a $3.12\,\text{m s}^{-2}$ **b** $20.0\,\text{N}$ **c** $3.06\,\text{m s}^{-1}$ **d** $2.45\,\text{m}$
9 $6.91\,\text{m s}^{-2}$

SKILLS CHECK 5C (page 57)

1 a $8\,\text{N s}$ **b** $1.8\,\text{N s}$ **c** $10\,\text{N s}$
2 a $96\,\text{N s}$ **b** $128\,\text{N s}$
3 a $60\,\text{N s}$ **b** $33\,\text{m s}^{-1}$
4 $400\,\text{N}$
5 $(-3.75\mathbf{i} - 3.75\mathbf{j})\,\text{N}$
6 $10\,\text{m s}^{-1}$
7 a $4.67\,\text{m s}^{-1}$ **b** $0.75\,\text{m s}^{-1}$ **c** $26\,\text{m s}^{-1}$
8 $3.69\,\text{m s}^{-1}, 1850\,\text{N}$
9 $300\,\text{m s}^{-1}, 45\,000\,\text{N}$
10 $6\,\text{m s}^{-1}, (4M)\,\text{N s}$

Exam Practice 5 (page 59)

1 a $2.45\,\text{m s}^{-2}$ **b** 0.25
2 a $\frac{21}{5}mg$ **c** 7
3 a $\frac{2}{5}g$ **b i** Same acceleration   **ii** Same tension   **c** $0.5\,\text{m}$
4 a $T = \frac{8}{5}mg$
5 a $1.75\,\text{m s}^{-1}$ **b** $0.75\,\text{N s}$
6 a $4.8\,\text{m s}^{-1}$ **b** $43\,200\,\text{N s}$ **c** $3600\,\text{N}$

SKILLS CHECK 6A (page 65)

1 a $27\,\text{N m acw}$ **b** $10\,\text{N m cw}$ **c** $0\,\text{N m}$
2 a $22\,\text{N m cw}$ **b** $170\,\text{N m cw}$ **c** $102\,\text{N m acw}$
3 a $18\,\text{N m}, 3.3\,\text{m}$ **b** $5.4\,\text{N m}, 21.6\,\text{N m}$ **c** $2.79\,\text{N m}, 1.79\,\text{N m}$
4 $1176\,\text{N}, 90\,\text{kg}$,
5 $20\,\text{kg}$, mass of children are taken as acting at a single point
6 $4.67\,\text{m}, 137\,\text{N}$
7 $0.23\,\text{m}, 314\,\text{N}, 157\,\text{N}; 112\,\text{kg}$
8 $3x\,\text{m}$

Exam Practice 6 (page 66)

1 $5.6\,\text{m}$
2 a $30\,\text{kg}$ **b** $3.6\,\text{kg}$ **c i** Weight acts at centre   **ii** Weight acts at A.
3 a Rod   **b** $2352\,\text{N}$ **c** $2058\,\text{N}$
4 a $\frac{5}{6}$
   b Weight acts through mid-point, AB remains a straight line.
5 a 0 **b** $3750\,\text{N}$
   c $3125\,\text{N}$ **d** $1.6\,\text{m}$

Practice exam paper (page 69)

NB Any numerical answer which has been obtained using $g = 9.8$ m s^{-2} should be given to two significant figures.

1 a $\mathbf{a} = \dfrac{\mathbf{v} - \mathbf{u}}{t} = \dfrac{(14\mathbf{i} - 19\mathbf{j}) - (-4\mathbf{i} + 5\mathbf{j})}{3} = 6\mathbf{i} - 8\mathbf{j}$

b $\mathbf{F} = 4\mathbf{a} = 24\mathbf{i} - 32\mathbf{j}$
$|\mathbf{F}| = \sqrt{(24^2 + 32^2)} = 40$ N

c $\tan \theta° = \frac{32}{24} \Rightarrow \theta = 53$

2 a

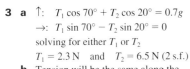

CLM: $4m \times 2u - 3m \times 2u = 3m \times v$

$v = \dfrac{2u}{3}$

Direction is opposite to original direction

b $I = 4m(0 - (-2u))$
$= 8mu$

3 a $\uparrow$: $T_1 \cos 70° + T_2 \cos 20° = 0.7g$
$\rightarrow$: $T_1 \sin 70° - T_2 \sin 20° = 0$
solving for either T_1 or T_2
$T_1 = 2.3$ N   and $T_2 = 6.5$ N (2 s.f.)

b Tension will be the same along the length of each rope.

4 a $\uparrow$: $T + 5T = 20g + 12g + 10g$
$5T = 35g = 343$ N   or 340 N (2 s.f.)

b $M(A)$, $10gx + 40g + 12g \times 3.5 = 35g \times 3$
$x = 2.3$
Distance between particles is $3.5 - 2.3 = 1.2$

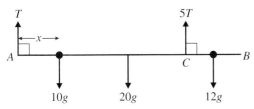

5 a When $t = 0$, $\mathbf{v} = \mathbf{i} - 3\mathbf{j}$
Speed $= |\mathbf{v}| = \sqrt{(1^2 + 3^2)} = \sqrt{10} = 3.16$ (in m s^{-1})

b Moving parallel to $\mathbf{i} - \mathbf{j}$ when
$(3t^2 - 4t + 1) = -(5t - 3)$
$3t^2 + t - 2 = 0$
$(3t - 2)(t + 1) = 0$
$t = \frac{2}{3}$ or -1
$t = \frac{2}{3}$

6 a $\uparrow$: $R = 40g - 150 \sin 30° = 317$
$\rightarrow$: $\mu R = 150 \cos 30°$
$\mu = 0.41$ (2 s.f.)

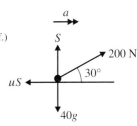

b $\uparrow$: $S = 40g - 200 \sin 30° = 292$
$\rightarrow$: $200 \cos 30° - \mu S = 40a$
$a = 1.3$ m s^{-2} (2 s.f.)

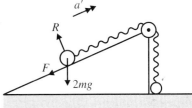

7 a $s_P = t + \frac{1}{2}2t^2 = t + t^2$
$s_Q = 16(t - 4) + \frac{1}{2}(t - 4)^2$
Need $s_Q > s_P$
i.e. $16(t - 4) + \frac{1}{2}(t - 4)^2 > t + t^2$
$32t - 128 + t^2 - 8t + 16 > 2t + 2t^2$
$0 > t^2 - 22t + 112$
$0 > (t - 8)(t - 14)$
$8 < t < 14$

b P overtakes Q after 14 s
when $t = 14$, $s_P = 14 + 14^2 = 210$ m

8 a

$3mg - T = 3ma$
$T - 2mg \sin \alpha° - F = 2ma$
$R = 2mg \cos \alpha°$
$F = \frac{11}{36} R = \frac{22}{39}mg$
Solving for a
$a = \frac{1}{3}g$

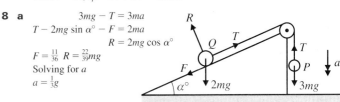

b Let V be the speed of P when it hits the ground.
$V^2 = 2 \times \dfrac{g}{3} \times 2 = \dfrac{4g}{3}$
After P hits the ground, T will be 0, so
$-2mg \sin \alpha° - F = 2ma'$, where a' is the deceleration of Q.
$a' = -\frac{2}{3}g$
$0^2 = \dfrac{4g}{3} - 2\dfrac{2g}{3}s \Rightarrow s = 1$
so distance from pulley
$= (6 - 2 - 1)$ m $= 3$ m

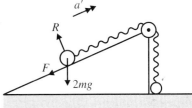

SINGLE USER LICENCE AGREEMENT FOR MECHANICS 1 FOR EDEXCEL CD-ROM
IMPORTANT: READ CAREFULLY

WARNING: BY OPENING THE PACKAGE YOU AGREE TO BE BOUND BY THE TERMS OF THE LICENCE AGREEMENT BELOW.

This is a legally binding agreement between You (the user or purchaser) and Pearson Education Limited. By retaining this licence, any software media or accompanying written materials or carrying out any of the permitted activities You agree to be bound by the terms of the licence agreement below.

If You do not agree to these terms then promptly return the entire publication (this licence and all software, written materials, packaging and any other components received with it) with Your sales receipt to Your supplier for a full refund.

YOU ARE PERMITTED TO:

- Use (load into temporary memory or permanent storage) a single copy of the software on only one computer at a time. If this computer is linked to a network then the software may only be used in a manner such that it is not accessible to other machines on the network.

- Transfer the software from one computer to another provided that you only use it on one computer at a time.

- Print a single copy of any PDF file from the CD-ROM for the sole use of the user.

YOU MAY NOT:

- Rent or lease the software or any part of the publication.

- Copy any part of the documentation, except where specifically indicated otherwise.

- Make copies of the software, other than for backup purposes.

- Reverse engineer, decompile or disassemble the software.

- Use the software on more than one computer at a time.

- Install the software on any networked computer in a way that could allow access to it from more than one machine on the network.

- Use the software in any way not specified above without the prior written consent of Pearson Education Limited.

- Print off multiple copies of any PDF file.

ONE COPY ONLY

This licence is for a single user copy of the software

PEARSON EDUCATION LIMITED RESERVES THE RIGHT TO TERMINATE THIS LICENCE BY WRITTEN NOTICE AND TO TAKE ACTION TO RECOVER ANY DAMAGES SUFFERED BY PEARSON EDUCATION LIMITED IF YOU BREACH ANY PROVISION OF THIS AGREEMENT.

Pearson Education Limited and/or its licensors own the software.
You only own the disk on which the software is supplied.

Pearson Education Limited warrants that the diskette or CD-ROM on which the software is supplied is free from defects in materials and workmanship under normal use for ninety (90) days from the date You receive it. This warranty is limited to You and is not transferable. Pearson Education Limited does not warrant that the functions of the software meet Your requirements or that the media is compatible with any computer system on which it is used or that the operation of the software will be unlimited or error free.

You assume responsibility for selecting the software to achieve Your intended results and for the installation of, the use of and the results obtained from the software. The entire liability of Pearson Education Limited and its suppliers and your only remedy shall be replacement free of charge of the components that do not meet this warranty.

This limited warranty is void if any damage has resulted from accident, abuse, misapplication, service or modification by someone other than Pearson Education Limited. In no event shall Pearson Education Limited or its suppliers be liable for any damages whatsoever arising out of installation of the software, even if advised of the possibility of such damages. Pearson Education Limited will not be liable for any loss or damage of any nature suffered by any party as a result of reliance upon or reproduction of or any errors in the content of the publication.

Pearson Education Limited does not limit its liability for death or personal injury caused by its negligence.

This licence agreement shall be governed by and interpreted and construed in accordance with English law.